EXPERT WITNESSES

Science, Medicine,

and the Practice of Law

CAROL A. G. JONES

Clarendon Press · Oxford
1994

Oxford University Press, Walton Street, Oxford OX2 6DP
Oxford New York Toronto
Delhi Bombay Calcutta Madras Karachi
Kuala Lumpur Singapore Hong Kong Tokyo
Nairobi Dar es Salaam Cape Town
Melbourne Auckland Madrid
and associated companies in
Berlin Ibadan

Oxford is a trade mark of Oxford University Press

Published in the United States
by Oxford University Press Inc., New York

British Library Cataloguing in Publication Data
Data available

Library of Congress Cataloging in Publication Data
Jones, Carol A. G.
Expert witnesses : science, medicine, and the practice of law /
Carol A.G. Jones
p. cm. — (Oxford socio-legal studies) Includes index.
1. Evidence, Expert—Great Britain. 2. Forensic sciences—
Great Britain. 3. Science and law. I. Title. II. Series.
KD7521.J66 1993 [344.10567] 345.41'067—dc20
93–14378

ISBN 0–19–825797–X

Set by Hope Services (Abingdon) Ltd.
Printed in Great Britain
on acid-free paper by
Biddles Ltd
Guildford & King's Lynn

For Pater

Acknowledgements

As with most academic work, this book is the result of innumerable interactions with different people. I am first indebted to Doreen McBarnet, who helped shape and focus my jumbled thoughts, and to the late Sir Rupert Cross, who patiently guided the sociological novice through the law of Evidence. I have also benefited enormously from the input of Barry Barnes and Maureen Cain. I owe a huge debt of thanks to my numerous respondents whose identity must remain anonymous, and to the court clerks, ushers, and office assistants who made the fieldwork part of this study both possible and pleasant. For their practical help and support with the final production I must also thank Joe Cavanagh, Georgia Duckworth, and Sarah Maidlow, as well as my editors at Oxford University Press, whose diligence and hard work is much appreciated.

Contents

1 Introduction

This is a book about expert witnesses. However, it is not simply a book about expert witnesses: it is also a book about law and science, their histories, their practices, the interaction between their bodies of knowledge, their discourses, and their professional communities. The relative absence of a book about the relationship between science and law is hardly surprising. Sociologists of knowledge have frequently pointed out that gaps in our knowledge betoken the wider structuring of knowledge in society. Law and science, like many of our powerful institutions, do not invite study.[1] They are complex and forbidding bodies of knowledge which intimidate the uninitiated. Together, they represent the official version of reality. This goes some way towards explaining why it is only relatively recently that sociologists and historians have mounted a constructivist critique of the cultural authority of law and science.[2] The idea that science and law might be historically specific social activities is still rather novel. Sociologists have been as prone as anyone else to seeing them not as cultural artefacts but as sacred commandments.[3] But if, as Clifford

[1] D. McBarnet, 'PreTrial Procedures and the Construction of Conviction', in P. Carlen (ed.), *The Sociology of Law*, Sociological Review Monograph 23 (Keele University, 1976), 176.

[2] For texts on law see e.g. D. McBarnet, *Conviction: Law, The State and the Construction of Justice* (Macmillan, Oxford, 1981); McBarnet, 'Pre-Trial Procedures'; M. McConville, A. Sanders, and R. Leng, *The Case for the Prosecution: Police Suspects and the Construction of Criminality* (Routledge, London, 1991); For texts on science see e.g. K. Knorr-Cetina and M. Mulkay (eds.), *Science Observed: Perspectives on the Social Study of Science*, Sage (London, 1983); M. Mulkay, *Science and the Sociology of Knowledge*, Allen & Unwin, London, 1979); B. Barnes and D. Bloor, 'Relativism, Rationalism and the Sociology of Knowledge', in M. Hollis and S. Lukes (eds.), *Rationality and Relativism* (Blackwell, Oxford, 1982); B. Barnes, *Scientific Knowledge and Sociological Theory* (Routledge, London, 1977); H. M. Collins and T. J. Pinch, 'The Construction of the Paranormal: Nothing Unscientific is Happening', in R. Wallis (ed.), *On the Margins of Science: The Social Construction of Rejected Knowledge*, Sociological Review Monograph 27 (Keele University, 1979); B. Latour and S. Woolgar, *Laboratory Life: The Social Construction of Scientific Facts* (Sage, Beverley Hills, Califf., 1979).

[3] S. Harding, *The Science Question in Feminism* (Open University Press, Milton Keynes, 1986), 39.

Geertz has put it, science is something hammered together in some place to some purpose by partisans and devotees, it can be subject to questions such as, why has it been built in the way that it has? 'If knowledge is made, its making can be looked into.'[4] Sociology of science has developed a strong programme of social constructionism, examining science as a social and cultural phenonomenon. This book aims to examine science *and* law in these terms. It aims to illuminate the constructedness of legal verdicts and the scientific expertise which underpins them. I aim to show, as Catherine Belsey puts it, 'the activity behind the image' of value-free science and neutral law and look at 'the process of its production' [5].

To do this is to follow in a particular tradition in the sociology of law. In her work on the construction of conviction, Doreen McBarnet has demonstrated how the judicial verdict—of guilt or innocence—is constructed out of the public eye. She demonstrates how adjectival law—the law of pre-trial procedures—shapes informal practice in the construction of conviction. Thus adjectival law is transformed from its lowly status as an aspect of law to an important step in our sociological understanding of the link between the law in the books and the law in practice. This book charts similar territory, taking as its focus the means by which the formal permissions of the law shape the ways in which scientific evidence is allied to the dynamics of advocacy.

A further explanation for the absence of a sociological text on expert evidence lies in the fact that it has not generally been seen as a social issue. When I first began to think about this subject in the late 1970s, I was warned that 'experts aren't interesting'. They were an unproblematic feature of life in general and of legal life in particular. Such nonchalant acceptance of social phenomena is a challenge to the sociological inquirer, for whom anything taken for granted immediately becomes suspect. Events during the 1980s and 1990s gradually served to

[4] C. Geertz, 'A Lab of One's Own' (review of N. Tuana, *Feminism and Science*, L. Sciebinger, *The Mind has no Sex? Women in the Origins of Modern Science*, and D. Haraway, *Primate Visions: Gender, Race and Nature in the World of Modern Science*), *New York Review* (8 Nov. 1990), 19.

[5] C. Belsey, *Critical Practice* (Methuen, London, 1980).

problematize expert evidence and make it an object of inquiry. First, experts were discredited in a number of high-profile criminal cases. In 1981, in what was to be the first of sixteen Court of Appeal cases, a Home Office expert, Dr Alan Clift, was said to have been discredited as a scientist and as a witness, leading to an unprecedented scrutiny of the cases he had handled since 1953. At least 1,500 cases were examined, 534 in some detail. Several other Home Office experts were also publicly criticized later in the decade. The cultural authority of science was also brought into question by objectors at several controversial planning inquiries, the most notable being the Windscale and the Sizewell Inquiries. These debated at length the environmental and safety implications of nuclear power and also raised questions about the ability of laypersons to challenge the scientific and legal establishments. One issue became which science the tribunals deemed acceptable and which they dismissed as fringe science. A further issue was how much scientific debate and controversy the inquiries were prepared to entertain. Too much debate imperilled the possibility of a decision altogether; too little threatened the legitimacy of the exercise in public consultation.

The credibility of expert witnesses was further dented before an international public in the Australian case of *R.* v. *Chamberlain* (the so-called 'Dingo Baby' case). How could teams of high-calibre scientists possibly have mistaken copper oxide dust for fetal blood? The release of the Guildford Four and the Birmingham Six, and the doubts raised by the inquiry of Sir John May into the convictions of the Maguire Seven, all generated similar levels of disbelief: how could impartial Home Office experts possibly mistake innocent contamination for contamination by the explosive nitro-glycerine? A Royal Commission on Criminal Justice, set up in England and Wales in 1991 in the wake of these miscarriages of justice, was to consider, amongst other things, the role of expert evidence in the criminal justice system. A report previously produced in Scotland had already recommended changes to the organization of forensic services in that legal system.[6]

[6] Report on *The Forensic Pathology Services in Scotland* (The McCluskey Bowen Report) (Crown Office, Edinburgh, 1975).

Forensic science also became an issue in the debate about public sector spending which characterized so much of political life in the 1980s. The criterion for deciding how forensic experts were used became not whether they were necessary in the interests of justice but whether they were economic, efficient, and effective. The Rayner Initiative generated a series of internal management consultancy reports throughout the 1990s, resulting in a remodelling of the forensic science services in England and Wales. This resulted in their becoming an executive agency from April 1991. The allegedly greater independence of this agency is an issue I examine in some detail at a later point in this book. A by-product of this high degree of internal management activity within the Home Office Forensic Science Service was a growing awareness of falling morale amongst its employees. Criticisms which hitherto would have been expressed only inside the Service itself began to surface in public. The Service was subjected to the critical scrutiny of the Parliamentary Home Affairs Committee. Experienced Home Office experts departed for life in the private sector, thereby forming for the first time a small but significant body of credible opposition which could and did challenge the findings of State forensic science.

In the 1980s, a growing public concern with environmental issues also detracted from the image of science as the benefactor of mankind. Traditionally, the pursuit of scientific progress has been billed as an emancipatory endeavour. In particular, scientific progress was credited with delivering the benefits of modernization. In the 1980s, this view began to give way in the face of an alternative world-view, in which living in harmony with nature became preferable to gaining mastery over it for the purposes of profit and exploitation. Science was criticized for developing oppressive rather than emancipatory technologies which served the interests of dominant groups. The selection and definition of problematics—deciding what phenomena in the world need explanation, and defining what is problematic about them—was said to bear the 'social fingerprints' of dominant groups.[7] The question of whether there

[7] Harding, *The Science Question*, 22.

could ever be such a thing as value-neutral science—traditionally a question confined to the esoteric world of philosophy—now became a part of an active and wider agenda of debate.

Cumulatively, this constellation of events reconstituted expert evidence as a social problem. In one sense, then, this book is a timely intervention in a growing debate about the place of scientific experts in society generally, and in the legal system in particular. Hopefully, it also fills a gap in our knowledge. It is the first comprehensive socio-legal analysis of experts in the legal system. It looks at experts as scientists, as witnesses, and as an aspect of advocacy. It looks at how lawyers use experts to construct stories, and at how they control and filter expert evidence at the pre-trial and trial stages. It also raises questions about images of science and their social uses and argues that, despite the legal and scientific dogma that the facts speak for themselves, in practice no one in the legal system leaves what they say to chance. I try to show that determining what the facts say is a process of persuasion in both the legal and the scientific communities. Legal and scientific facts are not given; they are highly pre fabricated. To look at how experts participate in the construction of facticity is to look at how they contribute to the apparent immutability of legal verdicts.

The view of science which has been co-opted by law is a caricature of science. Sometimes termed scientism, this view holds that scientific knowledge is the only objective truth, that science alone can observe the empirical reality of the world, and that, in making their observations, scientists employ a set of procedures and techniques which are neutral and hence provide an exact reading of the facts of reality. Scientists alone are held to render up this value-free reading of the world and, this being so, science claims to be a superior form of knowledge which transcends all other knowledge claims. As Porter has put it:

> Anxious lest to err is human, scientists have created a Frankenstein's monster, science as automaton. Too wise to present themselves as

supermen, they offer science as super-human, above the human comedy. Doing science is divine. Their enthusiasm for dehumanising science has been at best a mixed blessing, not least for scientists themselves. Luckily the humanities might help lift science off the horns of its self-inflicted dilemma. They can reveal how false the image is and also show that it is not a 'given' but the product of history. . . . If we think we must choose between Roundhead science and Cavalier subjectivity, between logic and irrationality, fact and fiction, nature and culture, we are impaled on false dichotomies. For science is itself a cultural product, employing rigorous man-made techniques to generate forms of knowledge which also mediate social needs and interests.[8]

The Roundhead account of what science is has been harnessed by the law to authorize powerfully its own version of events and render non-negotiable verdicts of evaluation and choice. How did this come about?

The Rise of the Fact/Value Distinction in Science

Creating and sustaining the distinction between fact and value has been one of the main activities of law and science over the last 200 years. The distinction between fact and value in science is mirrored by the distinction between fact and opinion in law. This distinction is critical to our understanding of how the expert witness role developed. We are told that this distinction underpins much of the law of evidence, that it particularly underwrites exclusionary rules regarding testimony based upon inferences or conclusions. We are also told that the one exception to this rule is the evidence of the expert witness. But we are never told why this all came about. It is as if the idea of distinguishing between fact and opinion was so overpowering in its logic that it simply fell from heaven into the vacant heads of the judiciary. In the history of science it is as if the facts leapt out of nature and struck the observer directly in the eye, on the head, or in the bath. In both law and science the role of the interpreter of fact is played down.

[8] R. Porter, 'White Coats in Vogue', *Times Higher Educational Supplement*, 5 Nov. 1982.

In science, we can locate the fact/value distinction in a particular historical epoch, what Harding has termed 'the moment of mythologizing'.[9] Harding locates the fact/value distinction in science in seventeenth-century England, when a number of other rigid dichotomies were formulated between, for example, nature and culture, objectivity and subjectivity, reason and emotion, form and content, theory and practice, individual and society, amateur and professional, truth and falsehood, appearance and reality. Such distinctions are part of what sociologists and others have termed the rise of modernity:

. . . the massive social and cultural changes which took place from the middle of the sixteenth century . . . it is consequently bound up with the analysis of industrial capitalist society as a revolutionary break with tradition and a social stability founded on a relatively stagnant agrarian civilisation. Modernity was about conquest—the imperial regulation of land, the discipline of the soul, and the creation of truth. . . . The question of postmodernism is a question about the possible limits of the process of modernisation'.[10]

Harding argues that the New Science of seventeenth-century England developed from the perspective of the skilled artisans of the time. It was, she argues, the mariners, shipbuilders, foundrymen, watchmakers, carpenters, and others whose task was the manipulation of instruments and raw materials who developed the method of learning by experimental observation. It was also the new science of engineering which pushed the law into its first major articulation of the place of scientific knowledge in legal proceedings.

The ascendancy of these skilled artisans corresponded with the breakdown of the feudal division of labour, the sweeping away of the old social order, and the rise of the bourgeoisie. The New Science in Puritan England developed within this context of social change, as part of a self-conscious political movement with radical, anti-authoritarian social goals: 'Full of self-confidence and enthusiasm, the various circles of scientists in England saw the political impulse of Puritanism and the

[9] Harding, *The Science Question*, 222.

[10] B. S. Turner (ed.), *Theories of Modernity and Postmodernity*, (Sage, London, 1991), 410.

struggles of emerging science as a single progressive dynamic in the shaping of postfeudalism. Science's progressiveness was perceived to lie not in method alone but in its mutually supportive relationship to progressive tendencies in the larger society'.[11] In the feudal world-view, change was seen as signifying corruption and decay; the New Science held change to be both possible and desirable; it had what Harding terms a humanitarian orientation, committed to the pursuit of projects for the public good. The New Science thus explicitly embraced the development of learning through the scientific method as a moral ethic. The idea that science is inherently emancipatory thus arose from a commitment to the social uses of science.

When did this change? When did science cease to embrace social values and begin to distance fact from value? According to Harding and van den Daele, the Restoration saw the replacement of Puritan progressiveness by absolutist rule, resulting in sanctions against the 'social programmes of science': 'Here we have the moment of mythologising. . . The end of the Puritan Revolution with the Restoration in 1660 also marked the end of the association between science and social, political or educational reform. . . . The political and social setting of the new science was fundamentally changed.'[12]

This division encouraged the notion that 'scientists as scientists were not to meddle in politics; political, economic, and social administrators were not to shape the cognitive direction of scientific inquiry'.[13] But, as Harding points out, such a division is hard to maintain, partly because those who hold political and economic power can and do settle the agenda for scientific research, partly because 'individual scientists and the scientific enterprise itself are social artifacts, and the selection and definition of what needs explaining can never be free of social dimensions'.[14] The institutionalization of science ensued in the period immediately after the Restoration with the founding of the Royal Society in 1662.[15]

Looked at this way, it is possible to identify what Harding calls the historical compromise which formed the key compo-

[11] Harding, *The Science Question*, 219. [12] Ibid. 223. [13] Ibid. 222.
[14] Ibid. [15] Ibid.

nents of modern science's commitment to value-neutrality: 'The claim that science is value-neutral was not arrived at through experimental observation; it was instead a statement of intent, designed to ensure the practice of science a niche in society rather than the emancipatory reform of that society.'[16] The fact/value distinction was a particular ideological device which accomplished this niche for scientists. It produced a self-serving version of science in which the facts speak for themselves without reference to their political and economic dimensions. Indeed, this version of scientific facts allows us to talk about them as if they were inert and passive, free of any controversy whatsoever. Scientists are seen as disinterested, open-minded, impartial, emotionally detached, supremely rational, and possessed of a scrupulous mental hygiene which transcends ideology, politics, and self-interest.[17] They can be trusted and believed. Their status as fountains of cultural authority has become so entrenched that we take it for granted. Likewise, the law has used the fact/value distinction to disguise the role of choice and evaluation in legal discourse. As a result, its pronouncements equally seem to be set down in tablets of stone.

The argument of men of science has been that 'science's logic and methodology, and the empirical core of scientific facts these produce, are totally immune from social influences; that logic and scientific method will in the long run winnow out the factual from the social in the results of scientific research'.[18] Law makes much the same claim, arguing that special legal tools make the separation of fact and opinion possible. Central to legal ideology is the claim to judicial neutrality, the doctrine of precedent which 'binds' judicial decisions, well-established principles, and the application of logic and reason.[19] Men of science and men of law have thus come to enjoy an unrivalled position as diviners of the facts. Their findings are held to be the result not of personal whim but of

[16] Ibid. 224.

[17] I. Cameron and D. Edge, *Scientific Images and their Social Uses* (Butterworths (Science in Context Series), London, 1979), 15.

[18] Harding, *The Science Question*, 40.

[19] A. Sachs, 'The Myth of Judicial Neutrality', in P. Carlen (ed.), *The Sociology of Law*, 1976, 110.

impersonal judgement. The version of the facts they reveal is thereby held to be free of normative relativism; the facts are simply the objective truth of what is and always has been.

For Weber of course the rise of rationality was initially a liberating force and Harding's account of the rise of the New Science supports this view. However, as Sachs has argued, the notion of impartiality is itself value-laden, and the doctrine of judicial neutrality is itself a myth.[20] The legal profession in England, he argues, created for itself an almost entirely fictional idea of legal history in which custom had been uninterrupted since time immemorial. In his view, judges see the future as a re-creation of the past instead of a re-creation of the future.[21] Rather than being liberating, law is thus inherently conservative, favouring those with an established position of power in society. The alliance between the abrogated form of the New Science and law produces a particularly virulent form of rational-legal domination.

Some Reservations

In order to develop my arguments, I have organized my text into several main sections, the first dealing with the history of the early modern and modern legal systems, the latter with developments in what I have called the late modern legal system. In making this distinction I am conscious of two things. The first is that by adopting this stylistic device I may seem to be attributing a certain unilinear progression to the law, but I hope it will become clear that such a result is not intended. Indeed, I suggest that an overarching concept of a uniform progression, marching, as Stone puts it, 'relentlessly through the centuries' in the direction of modernity, is itself a legacy of the ideology of Social Darwinism which has reached the limits of its legitimacy.[22] It reduces social diversity to a uniformity which has never existed in real life and is itself bequeathed to us by the ideology of modern science.[23]

[20] Ibid. 121. [21] Ibid. 122.

[22] L. Stone, *The Family, Sex and Marriage in England 1500–1800* (Penguin, London, 1977), 422. [23] Ibid.

My second reservation is that the terms early modern, modern, and late modern may be assigned a rather different meaning from the one intended. I do not, for example, intend these terms to be interpreted as part of the lexicon of the post-modernist debate, although part of my argument does focus on the promise of modernity. My interest in the legal system in the late twentieth century lies in how legal forms have shaped and constrained scientific expertise, by what means, and in what circumstances. How have the distinctions forged during the rise of modernity fared? How did the idea of a techno-legal rationality administered by experts take hold and how has the legal system resisted it? How has the ascendancy of reason over nature come to be challenged in the latter part of the twentieth century?

I also look at how what Larson calls the professional projects of law and science, formulated in the nineteenth century, have developed in the twentieth century.[24] I argue that the history of the expert witness cannot be understood outside this context of the competing professional bids for power in social decision-making. The history of law and science can be written in very different terms from the official version, and in the twentieth century many writers have sought to place scientific and legal developments in their socio-economic and historical context. The philosophy and sociology of science have gone a long way towards deconstructing the official version of science and its history, and have provided an alternative account of the way in which knowledge is acquired and incorporated. Like legal practice, science is subject to paradigm shifts during the course of which pre-existing ideas and practices are set in flux. If science appears to have been subject to closer scrutiny than law in this area, it is probably because sociologists have focused more upon the formation of scientific knowledge than they have upon legal knowledge. Few have looked at forensic science, with the result that the strong reifying tendencies of science and law have been allowed to combine without any critical analysis.

[24] M. Larson, *The Rise of Professionalism* (University of California Press, Berkeley, Calif., 1977).

The analysis I present here reinstates a couple of items which the law has erased from its own account of the history of expert witnesses. The first item to be reinstated is the place of power in the history of law and its procedures. Seemingly insignificant alterations to rules and procedures have gradually installed the judiciary in a position of control which is virtually unassailable. The fact/opinion rule was one part of the process which led to this control; so was moving the expert to the witness-box, and the creation of the new model jury. A second item to be reinstated is the interplay of social, political, and professional interests. The law deems these extra-legal; science deems them unscientific and therefore of no real account. This alone should give us every reason for thinking they may actually be of some consequence.

For a discipline which routinely talks in terms of precedent and prides itself on its long history, the law gives a curiously ahistorical account of itself. Scientists have fashioned their own history likewise, suggesting an unbroken and unproblematic progression from one stage to the next. More than this, their histories project an inevitable progression towards greater and greater rationality. So, just as legal history presents us with an early phase characterized by trial by battle, early forensic science is presented as a practice bound up with divination, ordeal, and alchemy. Standard legal histories chart a progression away from older forms of proof such as compurgation, battle, and ordeal, towards trial by reason, with decisions being reached on evidence presented before a judge and jury in open court. Legal historians argue, for example, that trial by jury had begun to replace the older forms of proof in civil cases by the fourteenth century. As evidence they cite an action of trespass in 1304 in which proof by battle was refused.[25] In fact, however, trial by battle actually remained an option in criminal cases until the nineteenth century. One is invited to believe that by the twentieth century law has progressed to the zenith of rationality. The law of evidence and procedure claims to be a logical system of established rules;

[25] Y. B. 32, 33 Edw. I (RS) 318, 320.

texts often refer to the rules of evidence in terms which suggest that their origin has been lost in the mists of time; if there ever was a human hand involved, it is no longer visible. Yet law is also seen as changing with the times, by which means it can take account of the individuality of cases and permit expeditious invention. English legal rhetoric portrays this combination of rigid fixity with practical flexibility as one of the main virtues of the common-law system. However, in order to demonstrate the progressive nature of law, legal writers have also insisted on the indisputability of dates by which certain rules had been developed. There are some problems finding out what happened in the period between precedents, but exactly how practices progressed towards their final determination is said to be mere detail. Finding the missing links is downgraded as the esoteric work of academics. All that it is really necessary to know is that the law moves progressively from precedent to precedent in an ever more rational and logical fashion. In these ahistorical, asociological accounts, problems are smoothed out in a vacuum by invisible enquirers. The human hand on the tiller of change is negated. To speak of rules being settled and procedures being established is to be beguiled by the law's own version of its development.

I attempt in this book to restore the ambiguities and contingencies of both scientific and legal practice and to restore the human face of the invisible enquirer of legal and scientific rhetoric. The experience of expert witnesses illustrates the limits of these scientistic and legalistic accounts of men of science and men of law, accounts which portray them as being above material and ideological influences. Their work is supposed to be devoid of human or personal input, their findings the value-free, uncorrupted testimony of the facts. When scientists fail to live up to this ideal in the law courts, it is often said that this is because expert evidence is science for specific legal purposes. It is applied science rather than pure science and is thus held to be tainted science. Moreover, because experts give their evidence in an adversarial setting, it is easy to blame any problems with their evidence upon the built-in partisanship of a two-party system. The dynamics of advocacy are not so much concerned with deconstructing the other side's case as with destroying it altogether. One of the ways of achieving this has been to destroy the expert's credibility, on the basis that a bad man inevitably produces bad science.

More than most, then, expert witnesses find their views exposed to what Wynne has called conditions of organized scepticism. When they fail to meet requirements of the ideal man of science, they are dismissed as charlatans, visionaries, or fringe experts. The fake and the partisan hired gun are the favourite bogymen of lawyers' stories about experts in court. Such stereotypes only make sense set against society's preferred model of the cool, objective, correct, impartial man of science. Lawyers help mould these stereotypes. They routinely require their scientific witnesses to mask the contingent nature of their conclusions and their methodologies. Scientists must become expert not only as scientists but as *witnesses*. The term 'expert witness' is thus nicely ambiguous in this respect.

How do experts balance the expectations of a man of science against the requirements of their man of law role? Part of this book concerns itself with the playing out of this opposition. But ultimately, I argue that this dichotomy is a false dichotomy, that the question itself proceeds on a false premiss. Science and law are often seen as qualitatively different kinds of activities, the one concerned with objective facts, the other with subjective values. They are seen as separate but parallel modes of truth finding, each employing empirical methods and distinctive modes of reasoning. I take issue with this account, arguing that law and science are in many respects rather similar social practices.

At one level this is a normative project, one which seeks to correct society's excessive respect for the authority of science and law and challenge their regimes of truth. It proceeds from the basis that the doctrine of scientific impartiality, like the doctrine of judicial neutrality, is itself a social construct. Using a terminology of objectivity, each discipline disguises the role of interest and choice in its findings. To cite Sachs, the notion of neutrality is one of the 'sacred tales of the present'.[26] But

[26] Sachs, 'The Myth of Judicial Neutrality', 127.

since the special quality of myth 'is not that it is false but that it is divinely true for those who believe it, but fairy tale for those who do not', to criticize the impartiality of law and science is not to accuse the believers of hypocrisy or dishonesty.[27] As Sachs points out, what is self-evident truth to the believers is a manifest distortion to the non-believer. Part of the sociologist's task is to understand how this belief in the self-evident nature of science and law came about, in order to jolt us out of that somnambulistic certainty with which we regard certain features of social life.[28]

This book is written from the point of view of a non-believer. It attempts to show how legality is constructed, how science is employed to reify judicial verdicts. These, I argue, are not the revealed truths of the past but matters of choice; their apparent neutrality and facticity is the result of considerable work carried on beyond the public view. This work makes the unknowable (what really happened) into a cut-and-dried, logically complete case, devoid of what McBarnet has termed the ambiguities which beset real life.[29] McBarnet has argued that pre-trial procedures help to construct guilt out of the public eye, displacing the presumption of innocence by a presumption of guilt and filtering the information which comes to trial to produce a conviction. This book examines one particular aspect of this process. It shows how the science underpinning cases is shaped by the pre-trial processes of case construction. I also aim to demonstrate how informal practice is shaped by the formal permissions of the law in respect of pre-trial procedures guiding the disclosure and exchange of information.

This text is thus written partly in the hope of persuading some of the believers to look again at the practice of science and law, and to reread the history of their relationship in the context of wider social forces. These include the impact of economics and politics on the shaping of science and technology as well as the rise of the State and increasing centralization of

[27] Ibid.

[28] K. Mannheim, *Ideology and Utopia* (Routledge, London, 1946), cited in Sachs, 'The Myth of Judicial Neutrality'.

[29] McBarnet, *Conviction.*

power. I will try to outline the struggles of law to come to terms both with the rapid expansion of science in the nineteenth century, and with the challenge posed by a rival body of knowledge whose claim to cultural supremacy threatened the foundations of the traditional legal paradigm. I focus on one particular branch of science—what we now term forensic science—since it is a very specific case of the co-optation of science for social purposes.

2 Science, Law, and the Rise of the Modern State

The History of the Forensic Sciences

The place of the expert witness in the legal system is the result of a specific ideological formation between science, law, and the State. To date, the history of the forensic sciences has tended to ignore this. The history of forensic medicine and forensic science has been written by lawyers, doctors, or scientists. One of their main aims seems to have been the identification of forensic firsts, that is, the first recorded occasion on which a particular type of scientific knowledge was used in legal proceedings. Sociologically, the interesting questions are how and when did such knowledges develop? When did experts become distinct specialists? Under what kinds of social, economic, and political conditions did they rise and thrive? How did they obtain widespread social currency? Does the place of expert evidence in the law tell us something about the rise of science and the stratification of knowledge in society more generally? How did society and politics shape the form of forensic investigation? To what extent did the law and politics shape knowledge?

By the sixteenth century, forensic medicine was becoming an institutionalized activity in several European countries, including predominantly Catholic countries such as France and Italy. The Roman Catholic Church needed medical experts to deal with 'false miracles and those produced by sorcerers' and to test beliefs such as 'bees don't sting virgins'.[1] The violent deaths of popes and kings in medieval Europe also provided a specific impetus for forensic medicine, with medical men becoming physicians to the royal courts. Indeed, many of those credited with writing the first texts on forensic science were

[1] E. H. Ackernecht, 'The Early History of Legal Medicine', in C. R. Burns (ed.), *Legacies in Law and Medicine* (Science History Publications, New York, 1977), 252.

physicians such as these. The court physicians played an important role in society. Safeguarding of the purity of the royal line they presided over the birth of the royal children; on occasion they also presided over their conception. Their place in society was a reflection of the social significance of lineage both among the royal family and among the aristocracy. Physicians were employed to advise on such matters as legitimacy, premature births, suspected abortion, proof of virginity or defloration, the theory of male impotence, and any other physical factor which might affect succession.

A concern with death and forms of dying was also a reflection of the ecclesiastical courts' emphasis upon torture as a form of proof, and the emphasis upon bodily punishments in a society still very much focused upon what Spierenberg has termed the 'spectacle of suffering'.[2] Forensic medicine is closely linked to the development of different modes of punishment in different types of societies. Immink places the birth of public punishment in the twelfth century, alongside the birth of public penal law. The origins of both lie, according to him, in the context of changing relationships of freedom and dependence in feudal society.[3] In England, twelfth-century courts were part of the aristocratic administration of local estates and manors. The establishment of the Curia Regis represented the beginnings of a centralized judicial system, whilst the reign of Henry II was considered a new age in judicial administration.[4] This trend continued through the thirteenth century with the demise of the older courts and the strengthening of the centralized system. The system of assize courts allowed central government to control the conduct of local government, exercised by judicial officers.[5]

The history of expert witnesses must also be set in the context of developments in the English legal system, particularly

[2] P. Spierenberg, *The Spectacle of Suffering* (Cambridge University Press, Cambridge, 1984).

[3] P. W. A. Immink, *La Liberté et la peine: Étude sur la transformation de la liberté et sur le développement du droit pénal public en occident avant le XIIe siècle* (Assen, 1978), cited in Spierenberg, *Spectacle of Suffering*.

[4] W. A. Holdsworth, *A History of English Law* (14 vols.), vol i (Methuen, London, 1956).

[5] Ibid. 284.

the emergence of the State as investigator and prosecutor in criminal proceedings. The early modern phase as I describe it roughly coincides with the rise of the modern State in England, the rise and demise of the Curia Regis. It is a time of localized justice, of distinct spiritual and secular jurisdictions and mixed civil/criminal courts. The twelfth and thirteenth centuries see the beginnings of a centralized legal system, with the rise of the King's Bench and the Court of Common Pleas. It is a time in which the legitimate exercise of coercive power begins increasingly to be monopolized by king and State. From the fourteenth century onwards there was a growing separation between the legislative, administrative, and judicial arms of government. The court of King's Bench became separated from the king's person in the fourteenth century. It was with the rise of stabilizing nation-states, ruled by a sovereign king, that freemen also became the subjects of a public penal law. Thus the Hammurabi code, which represents one of the earliest examples of the power to regulate the practice of physicians and surgeons,[6] was also an early attempt to establish that the power to punish lay in the hands of the sovereign. Spierenberg observes that, until this point was reached, free men settled their disputes amongst themselves. Relatives of an injured or dead person could claim compensation from his attacker. The detailed study of wounds was necessary to establish whether they were offensive, defensive, self-inflicted, and so on. By such means it was possible not only to establish whether death had been caused by another person, but also to link the assailant with the victim by, for example, establishing that a wound to the head exactly matched the dimensions of a particular sword or axe.

The absence of any other type of forensic science except forensic medicine in early legal history is due to a strict stratification of knowledge in which specialization was absent (though alchemy was well established). Cosman's study of medieval medical malpractice makes it clear that the law

[6] B. Spector, 'The Growth of Medicine and the Letter of the Law', in Burns (ed.), *Legacies in Law and Medicine*, 274.

regulated medicine as a profession.[7] It expected medical men to understand everything from the nature of wounds to signs of witchcraft. Female doctors were gradually excluded from decision-making processes through laws which increasingly consolidated the power of medical men to regulate their own work. Distinctions arose between different branches of the profession such as surgeons, apothecaries, and physicians. There were local licensing laws which enabled only authorized individuals to practise; in 1518 the Royal College of Physicians was established and it, too, had powers to regulate the licensing of practitioners. Spector tells us that this met with great opposition from both the universities and the Church but its powers were consolidated by an Act of Parliament in 1540.[8] Spector also states that 1532 saw the first criminal code in Europe which contained statutory provisions directing the taking of medical testimony in all cases where death was occasioned by violent means.[9]

In early English practice, the role of medical men was also shaped by the imperatives of the coroner's court. Coroners' courts in the twelfth century were not primarily concerned with death, as they are today. The task of the coroner was the preservation and protection of the king's property, usually in the form of land or money. His interest in death sprang from the possibility that the Crown could demand the forfeiture of the material estate of the dead person. The Crown had a right to take possession of the estate of persons who had committed suicide, for suicide was a crime and whoever committed suicide was counted a felon. Again, a detailed study of the nature of wounds arose from the need to establish how and when they were caused, whether they were fatal, and whether they could have been self-inflicted. Forbes gives as an example the death of Sir Edmundbury Godfrey in 1679:

the Attorney General asked whether it was possible, as rumoured, that Sir Edmundbury might have hanged himself and that his rela-

[7] M. P. Cosman, 'Medieval Medical Malpractice', in S. Jarcho (ed.), *Essays on the History of Medicine* (New York Academy of Medicine, Neale Watson Publications, New York, 1976), 288.

[8] Spector, 'The Growth of Medicine', 287.

[9] Ibid. 288.

tives had thereafter run him through with a sword to save his estate. This raised an important legal point; if Sir Edmundbury while of sound mind had committed suicide, he would be accounted a felon and all his possessions would be forfeit to the Crown. But Skillard was convinced that the cervical dislocation was caused by more violence than a suicidal hanging'.[10]

The coroner was more likely to be a political appointee than a legal or medical man. He required medical experts to carry out an inspection of the corpse in order to provide an account of the cause of death. The morgue was the site of this work. It was from the examination of cadavers that elementary anatomical knowledge was first gleaned. Just as the penitentiary was to become the laboratory of nineteenth-century criminology, so the mortuary supplied the experimental laboratory of forensic medicine. Medical men monitored executions on behalf of the State and investigated all forms of sudden death. It was through the examination of the corpses of executed criminals that the science of anatomy progressed. Until the late eighteenth century, physical punishment inflicted in public on behalf of the State took the form of torture, various forms of death, mutilation, and maiming. Such forms of punishment were integral to a number of *ancien régime* legal systems, and surgeons were required to confirm that the punishment had been exacted in appropriate degree to secure a just measure of pain. Nineteenth- and twentieth-century pathologists were required to certify death following State executions. Medical men were later to assume other State functions as inspectors and controllers of prisons, workhouses, hospitals, and asylums.

Forensic expertise thus has a long-established cognitive and practical interrelationship with the State. What kind of knowledge was available was in part determined by legal imperatives, in part by wider social and political forces. The need for experts to aid the detection and proof of witchcraft only made sense in the climate of political upheaval which beset Europe in the Middle Ages. Other early court experts and special jurors were called to give evidence about the adulteration of

[10] T. R. Forbes, *Surgeons at the Bailey* (Yale University Press, New Haven, Conn., 1985), 47.

food, the making of false coins or promissory notes; guild-masters were required, as part of their social duties, to inform the law about quacks and instances of malpractice within the craft and skilled trades. The importance of shipping and sea-trade, coupled with the need for military and naval might, led others to become expert in skills of navigation, shipbuilding, cartography, and the construction of harbours. The need for nautical knowledge in Admiralty litigation led to the setting up of an institutional link between the law courts and the Brethren of Trinity House. Trinity House provides one of the few early examples of an institutional reservoir of expertise established outside the law courts. Unlike medical experts, whose main client was the State, the Brethren of Trinity House served both the State and private clients engaged in seagoing trades.

It took English forensic medical experts much longer than their Continental counterparts to establish themselves as a profession independent from the courts, with their own professional accrediting institutions. This had at least two effects: (1) their skills were channelled into areas of interest to the State, such as the classification of wounds, explanations for barrenness or impotence, and so forth; and (2) their training as forensic medical experts—in the law and the relevance of legal concepts—never matched that of their Continental colleagues, who early on had set up their own institutions which could train experts both in science and in law. This provided Continental experts with something English forensic medical experts lacked—an independent professional power base from which to reflect upon their own practice and that of the courts.

The History of The Expert Witness Role

My exploration of the early modern period is generally confined to the emergence of the witness role, changes in the form of trial procedure, and the growing demarcation between juror and witness, layperson and specialist. It was in this period that these distinctions were consolidated and the modern form of trial by jury formulated, that is, a jury composed of persons chosen at random from the community and having no prior acquaintance with the accused or the facts of the case. Before this, there had been little, if any, distinction between jurors and witnesses. The point at which juries ceased to be witnesses and became what Stephens terms 'judges of evidence given by others' is of critical importance to the development of the expert witness role.[11] The history of expert witnesses begins with this major structural change from quasi-judicial to witness status. Only in relatively recent times has the proper role for experts been construed as that of witness rather than judge or juror. In order to understand the significance of the expert witness role, it is therefore necessary to place it in the context of these structural shifts in the nature of the jury. Trial by a community of witnesses-cum-jurors was the main forerunner of trial by jury. It was not abolished as a form of trial until 1833 and was used in cases in 1708, 1799, and 1824. The early modern jury was chosen for its direct knowledge of the details of the case in hand. For the most part, witnesses were wholly unnecessary: 'When they appeared the jury could disregard what they said; and should, if it were not accordant with what they knew.'[12] So even where informing witnesses were called, there was no compulsion for jurors to take any notice of them. Indeed, it was perfectly possible and routine for trials to proceed without any witnesses at all.[13] The jurors *were* the witnesses. Only when the dual role was distilled into two separate roles (witness and juror) do we see the development of a distinct witness role in English courts of law.

The practice of community juries appears to have been a consolidation of customary and State law.[14] The original intent appears to have been to gather information about land and land-related matters. The jurors were sent around the country to collect information and report back. Their task was to supply facts from their own private knowledge and/or information

[11] Sir J. F. Stephens, *A History of the Criminal Law of England*, vols. i and ii (Burt Franklin, New York, 1883).

[12] J. B. Thayer, *Treatise on Evidence at Common Law* (Rothman Reprints, New York, 1969).

[13] YB 14 Hen. VII 29, 4.

[14] Holdsworth, *History of English Law*, i. 312.

which they had gathered from their investigations. On this basis, they then gave a verdict from their collective knowledge. The earliest recorded juries appear to have been employed in this way to discover and present facts in answer to inquiries addressed to them by the king. In this sense, the jury was one means of establishing State control under a single sovereign. The inquiries of the jury gradually extended beyond land-related questions to the discovery of suspected criminals. The task of the jury of presentment was to discover and present to the king's officials persons suspected of serious crimes. This kind of jury could number anything from twenty-four to eighty-four persons.[15] It heard no evidence. It consisted of whoever could or would support the story of one side against the other. Jurors were witnesses and investigators.[16] A jury from the same neighbourhood made this process trial by one's peers. One's peers might also be fellow craftsmen or tradesmen, persons within the same occupation, and/or persons of the same nationality. The community of witnesses could be any of these. For example, Thayer refers to a case in 1303[17] when a jury of 'goldsmiths and aldermen' was summoned. Likewise in 1280 a jury of 'Florentine merchants living in London' was summoned to try an act committed in Florence[18] and in 1351 and 1394 a jury was summoned comprising 'experts and men of particular trades, like the London juries of cooks and fishmongers, where one was accused of selling bad food'.[19]

[15] Thayer, *Treatise*, 85, and nn. 1 and 2. Thayer cites examples of early cases in which the number of jurors varied from sixty-six to four, five, or six. In 1199 there was a jury of nine (Cur. Reg. ii. 114). In Bracton's Note Book, at dates between 1217 and 1219, there are juries of nine, thirty-six, and forty.

[16] See YB 16 Edw. III (RS) ii. 16; Co. Litt. 295; Black. Comm. iii. 343. Dyer refers to the case of *Thorne* v. *Rolfe* (1560) being decided in this way.

[17] Pike, Hist. Crim. i. 198–200, 207, 208, 466.

[18] Pl. Ab. 201, col. 2. See also *Sherley's Case* 2 Dyer 114b; 3 and 4 P & M 315; also 2 Hawk. PC 591; *Symons* v. *Spinosa* 3 Dyer 357b, Easter Term 19 Eliz. 802; Brooks New Cases, 140 ER 908; 3 Dyer 304a, 13 and 14 Eliz. 683, 51.

[19] Ryley, *Memorials of London Life in the Thirteenth, Fourteenth and Fifteenth Centuries* (Longman & Green, London, 1863), 206 and 536.

Jurors with Special Knowledge

On occasion, the jury of fellow countrymen or neighbours needed the assistance of persons with special knowledge. In this situation, a jury would comprise ordinary persons with knowledge of the parties and the circumstances of the case, and specialist jurors to investigate certain circumstances and report back to the rest of the jury. Their combined knowledge supplied the basis of their verdict. This practice of calling a mixed jury applied equally to communities of skilled professionals or tradesmen, though on occasion such a jury would be composed entirely of persons with inside knowledge of the trade. A jury might comprise both specialist witnesses and community witnesses with the former 'supplying to the others their more exact information'.[20] In Thayer's terms:

> What we call the 'special jury' seems always to have been used. It was a natural result of the principle that those were to be summoned who could best tell the fact, the veritatem rei. And so we read that in 1645–6, in the King's Bench, 'The court was moved that a jury of merchants might be retained to try an issue between two merchants, touching merchants' affairs, and it was granted, because it was conceived they might have better knowledge of the matters in difference which were to be tried than others could who were not of that profession'.[21]

Riley refers to a number of fourteenth-century examples of special juries, summoned from trades or craftsmen to decide questions involving their craft.[22] Special juries were customary in the City of London to decide trade disputes. Records show that the mayor summoned the special jury from the relevant trade, while offenders were brought before the mayor by the supervisors of the guilds. Once the jury had reached its verdict, the

[20] Thayer, *Treatise*, 100. See also Br. NB iii case 1041; case 1717; case 1919. Also Holdsworth, *History of English Law*, i. 333; Bracton fo. 185b.

[21] Thayer, *Treatise*, 94–5; Lilley's Practical reg. ii. 154.

[22] See e.g. the Elizabethan case of *Hunter against Moone* 5 Eliz. Cap. 4 Noy. 133, where haberdashers and feltmakers deposed that someone trading as a dyer had not been apprenticed as one. See also 88 ER, KB 165; *The King against Kiffin* Cro. Eliz. 475; *Goodright* v. *Mist* 10 Geo. I 8 Mod. 248; also 3 Geo. II, c. 25 s. 15; 1733 Geo. II, c. 37 cap. xxxvii.

mayor passed sentence. Thus Learned Hand writes that in the developing urban community, 'where alone for the most part questions involving special skills would come up, the practice was well established in the fourteenth century of having the issue actually decided by people specially qualified'.[23]

Special Juries in Criminal Cases

Special juries were not confined to disputes within or between trades, nor to the sphere of civil law only. This practice appears to have been common to both English and French criminal law. Ackernecht provides an example of a jury of midwives from French law, in a case of pretended loss of virginity through rape:

> We, Jeanne de Mon and Jeanne Verguire and Béatrice Laurade, from the parish of Espoire in Béarn, matrons and midwives examined and approved, verify to whom it may concern that by order of the Judge of Espere, we, the undersigned matrons, have found, visited and seen, on May 15, 1545, Mariette de Garrigues, age 15, and the said Mariette said to have been raped and deflowered and devirginized. Therefore, we the undersigned midwives have examined and observed everything in the light of three candles, touched with our hands and examined with our eyes, and turned over with our fingers. And we found that neither was the vulva deformed nor the caruncu-lae displaced, not the labia minor distended, nor the perineum wrinkled, nor the internal orifice of the uterus opened, nor the cervix uteri split, nor the pubic hair bent, nor the hymen displaced, nor the breasts wilted, nor the margin of the great labia changed, nor the vagina enlarged, nor the membrane that connects the carunculae returned, nor the pubis broken, nor the clitoris in any way damaged. All this we, the above mentioned midwives, state as our report and direct judgment.'[24]

During the late eighteenth century, Forbes notes, there were seventy-three trials for sexual offences noted in the Old Bailey

[23] Judge Learned Hand, 'Historical and Practical Considerations Regarding Expert Testimony', *Harvard Law Review*, 15 (1901), 40. See also *Hunter against Moone*; Thayer, *Treatise*, 96–7; *The King against Burridge* 8 Mod. 230 Easter Term, 10 Geo. I, BR, 88 ER, KB 165.

[24] Ackernecht, 'History of Legal Medicine', 267.

records; seventy-one of the victims were females; over half were under the age of 11. He provides an English example of the alleged victim being examined by a midwife, apothecary, or medical man:

> 'I am an apothecary and a surgeon; I examined the prosecutrix.'
> 'What did you observe?'
> 'I observed she had a very great discharge, of a bad colour, yellow.'
> 'What do you mean, the foul disorder?'
> 'Yes.'
> 'Was there a great laceration of the parts?'
> No, I did not observe anything of that.'
> 'Did you examine the man?'
> 'He seemed to have no disorder at all, there appeared nothing.'
> Court to Jury: 'Gentlemen, then there is an end of this business, indeed there was an end of it before. Not Guilty.'[25]

Forbes also notes another case of a man being brought to trial for 'Ravishing and Carnally Knowing' an 11-year-old girl in 1735; she was examined by a midwife who found signs of abuse.[26] In the case of *R.* v. *Anne Wycherly*[27] a jury of married women was empanelled to determine if a convicted prisoner was with child. In another case, a girl of 7 who had allegedly been raped by a 60-year-old Scottish man was examined by 'divers women, matrons and a surgeon.'[28] The accused was found guilty, though the court 'doubted of rape in so tender a child. But if she had been of nine years and more, it would have been otherwise.' A second case from the same session speaks of the inspection of a girl of 9, to establish penetration in the case against Martin Page, servant, for the rape of the daughter of his master 'Wood, the Keeper of St. John's Head Tavern in Fleet Street'.[29]

Special Juries in Cases of Disputed Legitimacy

This kind of special jury was the jury *de ventre inspiciendo*, described by Bracton in the thirteenth century and referred to

[25] Forbes, *Surgeons at the Bailey*, 88.
[26] Ibid. 89.
[27] *R.* v. *Anne Wycherly* Cro. Jac. 541; Co. Litt. 8.
[28] 3 Dyer 304a, 13 and 14 Eliz. 683, 51.
[29] Easter 9, Car.

as the peculiar procedure for the empanelling of such a jury in cases where, in matter of inheritance, a woman falsely claimed pregnancy in order to disinherit another claimant.[30] The writ was issued 'to prevent such premeditated fraud' and involved taking 'lawful knights and lawful women' to see the woman involved and carefully examine her breasts and abdomen to see if she were pregnant. If the panel of women did find her pregnant, they were to inquire closely as to the time of conception, 'how, when and where . . . and at what time she believes she is to give birth'. On giving birth, it would then be possible to compute whether or not the child could be the child of a deceased husband:

Some say, though others are of a contrary opinion, that the woman cannot exceed the gestation period by a single day, even where the issue dies in utero or turns into a monster, the risk falling on the mother, but may anticipate the time of birth and deliver prematurely. If such a monster or prodigy is born it will not be reckoned as a child nor taken into account with respect to succession. But issue having more or fewer than the usual number of members, as six fingers on one hand, or no more than four, if in other respects he appears to be effectively human, shall be considered as a child with respect to succession . . . [where] there is a strong presumption,

[30] Bracton, De Leg. Lib. ii fo. 69. Bracton, *On The Laws and Customs of England*, trans. and ed. S. Thorne (The Belknap Press, Cambridge, Mass., 1968), 201–2. Translated by Thorne the writ runs as follows: 'The king to the sheriff, greeting. We command you, setting aside all delays and impediments, to cause to come before you and before the keepers of the pleas of our crown in your full court A. who was the wife of B. and who claims to be pregnant. And before the said keepers cause her to be examined by lawful and discreet women through whom the truth may be better known, and let the same women carefully examine her by feeling her breasts and abdomen and in every whereby they may best ascertain whether she is pregnant or not. And if the same women and keepers discover that she is pregnant, or if they are in doubt, then let them lodge her in our castle, such a one, in such a way that no maid who may be pregnant nor any other who may be suspected of contriving a deception has access to her, and let her remain in the castle until the question of her offspring can be settled. And provide that in that castle she may be separately guarded, lest any deception respecting her offspring occur while she is in custody. We have commanded such a constable to admit her to the aforesaid castle.'

The writ to the constable says that two or three women are charged with the matter and 'If they wish to examine her everyday, permit them to do so once a day . . . And let the same be done if she claims pregnancy in her husband's lifetime. If after his death she seeks, to the exheredation of the true heir, to be put in possession in the name of her unborn child, let this writ for proving the deception issue at the complaint of the heir.'

because of the interval of time or the distance between them, that the husband did not father such child, such issue will never be made legitimate, whether the husband avows it or not.'[31]

In *Willoughby's Case* it was said that a panel of women such as this 'shall see the woman whose condition is to be examined by certain specified methods, iquibus inde melius possit certiorari utrum proegnans sit necri'.[32] Sir Francis Willoughby died: 'seized of a great Inheritance, having five daughters, whereof the eldest was married to Percival Willoughby and not any sons. And the said Sir Francis, leaving his wife Dorothy, who at the time of his death pretended herself to be with Child by Sir Francis, which, if it were a Son, all the five sisters should thereby lose the Inheritance descended unto them'.[33] Percival and his wife Bridget obtained a writ *de ventre inspiciendo* from the Court of Chancery. The midwives reported Dorothy to be twenty weeks pregnant; she was kept under surveillance until the child's birth, to determine whether it be male or female, and eventually she produced a daughter.

In the case of *Alsop* v. *Bowtrell* (1619),[34] a jury of two doctors and five women and a midwife testified that a child born forty weeks after the death of the putative father could indeed be his legitimate child:

and this being proved [misuse of the mother by the father-in-law after the husband's death and before the birth] and this misusage by five women of good credit and two doctors of physic viz. Sir William Baddy and Dr. Munford, and one Chamerlaine (who was a physician and in the nature of a midwife), upon their oath, they affirming that the child came in time convenient to be the daughter of the party who died [the court holds that] it may well be as the physicians had affirmed that the said Elizabeth, who was born forty weeks and more after the death of the said Edmund Andrews, might well be the daughter of the said Edmund.[35]

[31] Ibid.
[32] *Willoughby's Case* (1597) Cro. Eliz 566; 78 ER.
[33] Grimston 1683 1, 56, cited in Forbes, *Surgeons at the Bailey*, 198.
[34] *Alsop* v. *Bowtrell* (1619) Cro. Jac. 541; 79 ER.
[35] Ibid.

The Decline of The Special Jury and the Demarcation of Juror and Witness Roles

In Holdsworth's view special juries were 'perhaps the ancestors both of the modern special jury and the modern expert witness. They partake of the character of both.'[36] The question must arise as to why and how it was that the institution was allowed to fall into decline. A good part of the answer to this question may be found in the development of the new model lay jury and the rules of evidence which came to define its province. By the fourteenth century rules had begun to develop which defined the conduct of jurors but there was still no distinct witness role, and hence no rules governing the conduct of witnesses. The rules shaping the conduct of the jury specified that a jury must stay together until it had done its duty.[37] It could neither eat nor drink until it had delivered its verdict.[38] For a time, jury decisions became very unstable. If the jury defied the judiciary, its verdict was simply referred to the jury of attaint. Whenever this second body disagreed with the finding of the original jury, the latter was deemed corrupt or biased. The alternative approach was simply to lock up the jury until it brought in the verdict the judge required.[39] For example, the jury in *Throckmorton's Case* in 1554 was imprisoned for refusing to convict the accused. These powers were still being used in 1664.[40] In 1665, however, it was decided that punishment should only be meted out to corrupt jurors.[41] The

[36] Holdsworth, *History of English Law*, i. 333.

[37] YB 24 Edw. III Hilary Term pl. 10. See also YB 13, 21, 22 Edw. I (RS) 272; 3, 4, Edw. II (SS) 188; YB 41 Edw. III Michaelmas Term; YB 12 Edw. III 31; s.c. 41 Ass. ii, Hale, Pl. Cr. ii. 297.

[38] YB 13, 21, 22 Edw. I (RS) 272; 3, 4 Edw. II (SS) 188.

[39] See Thayer, *Treatise*, 88–9 n. 4; Hen. VIII c. 4.

[40] See Thayer, *Treatise*, 164, 166; *Leach's Case* (1666) Th. Raym. 98.

[41] *R.* v. *Wagstaffe* (1665) Hale PC ii. 312–3, 6 ST 992. See also St. 5 Eliz. c. 9, s. 6; YB 5 Hen. VIII 8; YB 11 and 12 Edw. III 338; YB 12 and 13 Edw. III 44; 12 Ass. 34, 12; s.c. Fitz. Ab. Challenge 9; 23 Ass. 11; Thayer, *Treatise*, 137; YB 14 Hen. VII 29, 4. Holdsworth writes that it was not until Tudor times that 'the tide had turned in favour of the witness.' (Holdsworth, *History of English Law*, i. 334–5). Until this point, the law actively discouraged the use of witnesses. A statute of 1562–3 compelled witnesses to appear before the court of the Star Chamber, but witnesses were still only occasionally called in ordinary cases (see *Babbington* v. *Venor* (1465) Long. Quint. Edw. IV 58; s.c. YB 5 Edw. IV 51, 24). In 1607, in a proclamation drawn up by Bacon,

real turning-point came in *Bushell's Case* in 1670. According to Holdsworth, it was the argument of Vaughan in *Bushell's Case* 'which finally fixed the law' on the issue[42] and effectively put an end to the practice of locking up dissident juries.

Bushell's Case concerned a trial of Quakers in which the jury doggedly stuck to its verdict of not guilty. Quakers were an unpopular group with certain sections of seventeenth-century society and suffered a high degree of persecution. So where, as in this case, the jury refused to convict, it was kept without food and water for three days until it brought in the verdict required by the Recorder. In reviewing the case, Vaughan argued that punishing juries in this way was illogical. The fact that jurors differed from the judge made perfect sense, since they had private knowledge of the facts which they were entitled and required to use. The judge's knowledge, by comparison, was incomplete. He was only aware of the facts which had been produced in court: 'For the better and greater part of the evidence may be unknown to [the judge]; and this may happen in most cases, and often doth, as in Graves and Short's case.'[43] To insist that the jury must agree with the judge was therefore nonsense. Vaughan identified the route to controlling jury verdicts. If the jury's power lay in its private knowledge, the means to reduce its power was to reduce its private knowledge. By replacing a knowledgeable jury with a jury ignorant of the facts, it became necessary to call the evidence of witnesses. The judiciary would therefore know all that the jury knew:

> accordingly the old doctrine of their going on private knowledge began more and more to give way. The jury were told that if any of them knew anything relating to the case, they ought to state it publicly in court. This long lay in the shape of a moral duty of the jurors, not enforceable; but after a time it was enforced, and the court assumed that, in general, nothing was known to the jury except what was publicly stated in court . . . adding to this . . . what they

juries were still required to decide of their own private knowledge (J. Spedding, *The Philosophical Works of Francis Bacon* (Longman, London, 1857); Holdsworth, *History of English Law*, i. 33).

[42] Holdsworth, *History of English Law*, i. 345; (1607) Vaughan's Rep. 135.

[43] *Groves and Short's Case* Cro. Eliz. 616.

were legally supposed to know and what was known to everybody. This brought matters down to the state of things in which we are now living. The jury became merely judges upon the evidence.[44]

Thayer writes, 'as we see, not yet has the jury lost its old character as being in itself a body of witnesses; indeed, it is this character . . . that make[s] one of the chief pillars upon which Vaughan's great judgement rests. This double character of the jury was no novelty.'[45]

Following *Bushell's Case* it became possible both to order a retrial where the first produced an unsatisfactory verdict and to punish jurors who reached verdicts against the weight of the evidence. In the sixteenth century, the Star Chamber was apt to treat any verdict of acquittal which it considered to be against the weight of evidence as corrupt, and duly punished corrupt jurors by fine or imprisonment. The granting of a new trial provided another means of obtaining the required verdict. The first reported case of a new trial being granted was a civil case, *Wood* v. *Gunston*, in 1655 on the ground of excessive damages.[46] From the mid-seventeenth century onwards the jury's role became gradually more distinct from that of the witness. In the case of *Bennet* v. *Hundred of Hartford* in 1650 it was said that 'If a jury give a verdict of their own knowledge they ought to tell the court so that they may be sworn as witnesses . . . the fair way is to tell the court before they are sworn that they have evidence to give.'[47] In 1702, in the case of *Powys* v. *Gould*,[48] this view was reiterated. By 1816, the new paradigm of jury and witness roles had become well established. In the case of *R* v. *Sutton*, for example, the fact that a judge had directed the jury to reach a verdict based on their own knowledge was thought to be grounds for a new trial.[49] Lord Ellenborough expressed the view that 'a judge who should tell jurors to consider as evidence their own acquaintance with matters in dispute would misdirect them'. The true qualifi-

[44] Thayer, *Treatise*, 139. [45] Ibid. 168–9, 179.

[46] *Wood* v. *Gunston* (1655) Style 466, 1655, 82 ER 867. But see also Thayer, Treatise, 171 for a case in 1699, *Argent* v. *Sir Marmaduke Darrell.*

[47] Style 233.

[48] Salk 405, 1702; 7 Mod. 1; s.c. Anon. 1 Salk 405; Holt 404.

[49] 4 M. & S. 532.

cation for a juror had thus become 'exactly the reverse of that which it was when juries were first instituted. In order to give an impartial verdict he should enter the box altogether uninformed on the issue which he will have to decide.'[50]

Bushell's Case had laid clear the means of undermining the jury's power by reducing its old witness character. It was this power which had enabled it to oppose the Bench. 'By the end of another century, it would be gone.'[51] The modern form of the jury was therefore a complete reversal of the original form. Since the object of change had been to deprive jurors of their private knowledge base, experts had no place in this new jury. How was the jury to understand what special types of fact meant? How was it to relate them to all the other facts in the case? The jury needed to be supplied with such information by persons with the requisite knowledge and experience. The information could only be supplied by an expert giving evidence from the witness box. But the expert's testimony to facts and their interpretation was private knowledge and, as such, it posed a potential threat to the judiciary's newly acquired control.

In sum, the law needed the opinion evidence of expert witnesses but had restricted the province of witnesses to evidence of fact because to do otherwise would be to let in a threat to judicial control. To get itself out of this quandary, it devised a deceptively simple solution: it made expert witnesses an exception to the rule forbidding witnesses to give opinion evidence. From this point onwards experts became special sorts of witnesses. Only experts could give evidence of opinion. This is now the main legal rule about expert evidence. The new model jury and the fact/opinion distinction thus originated in judicial strategy to obtain greater control over trial outcomes. It was ironic, then, that having stripped the expert of his decision-making power by making him a witness, the law was forced to restore it to him by making him a special witness whose power flowed from his specialist knowledge.

By making experts into witnesses, the law placed them

[50] Pike, Hist. Crim. ii. 369; Thayer, *Treatise*, 170; *R.* v. *Sutton* 4 M. & S. 532.
[51] Thayer, *Treatise*, 169.

within the ambit of judicial control. Indeed, the judiciary appointed itself the protector of the jury's role since, by extension, this was part of the judicial province. It thus became necessary to protect it against experts. A strong judicial interest in claiming and maintaining control over legal proceedings thus played a significant part in determining the boundaries of the expert witness role. In the following chapters I aim to show how past and present legal practice is taken up with policing this expert/judicial boundary.

3 Court Experts

One of the problems with legal history is that it presents itself as a unilinear progression from one form to another. The narrative of a book tends to impose the same linearity. However, the history of expert witnesses fails to fit into this mould. The expert's witness role not only replaced the expert's special jury role, it also replaced the role of court expert. All three roles ran in parallel and still do. We have seen something of the expert's special juror role. What did this other role, court expert, consist of? The court expert role was quasi-judicial, the expert sitting at the right hand of the judge who had requested his appointment. His advice was given in the form of his conclusions upon the issues raised and was given without any form of cross-examination. The earliest reference to the practice of using court experts can be found in 1345, when surgeons were called to rule whether a wound was fresh. In 1353 there is reference to another case where surgeons were called to rule whether a wound was mayhem. In the case of *R.* v. *Coningsmark* in 1682, a surgeon was asked his opinion upon the nature of the bullet wounds and cause of death.[1]

Court experts were frequently doctors but this was not always the case. In 1554, in the case of *Buckley* v. *Rice Thomas* the judges referred to an earlier case (7 Hen VI) in 1429 in which the court listened to one Huls 'as men that were not above being instructed and made wiser by him'.[2] The issue on which they needed to be made wiser was the meaning of the Latin word *licet.* The court's Latin was said to have halted a little. Justice Staunford argued that the court should follow its predecessors who 'when they were in doubt about the meaning of any Latin words [inquired] how those that were skilled in the study thereof took them and pursued their construction'. As a precedent, he cited another case (9 Hen VII), where a

[1] *R.* v. *Coningsmark* (1682); *R.* v. *Ferrers* (1758) 1 Burr, 97 ER 483, 522.
[2] *Buckley* v. *Rice Thomas* (1554) I Plowd. 118, 75 ER, KB 182.

dispute had arisen over the meaning of the expression 'fine gold': the court had asked grammarians and others 'that best understood them' to interpret the words 'as grammar warrants and allows'. Mr Justice Saunders's response to this idea was as follows: 'And first I grant that if matters arise in our law which concern other sciences or faculties, we commonly apply for the aid of that science or faculty which it concerns, which is an honourable and commendable thing in our law. For thereby it appears that we do not despise all other sciences but our own, but we do approve of them and encourage them as things worthy of commendation.'[3] Courts appear to have been quite content to use a court expert or expert advisers to assist in the proper construction to be put on the wording of business and commercial papers, where the ordinary meaning of such words was inappropriate.[4] In 1702, in the case of *Buller* v. *Cripps*,[5] Lord Holt sought out the opinion of London merchants as to the effect of refusing promissory notes.[6] In 1753, in the case of *Fearon* v. *Bowers*,[7] and in 1755 in the case of *Ekins* v. *Macklish*,[8] Lord Hardwicke ruled in accordance with the views of merchants he had summoned to assist him in his judgment. A case from 1494 concerning the construction of a bond suggests that this practice was long-standing.[9]

In March 1665 there is reference to the appearance of a Dr Browne of Norwich in a case of witchcraft at Bury St Edmunds. Dr Browne was an eminent physician, 'a person of great knowledge, Dr Browne of Norwich, none other than Sir Thomas Browne, the physician philosopher who wrote *Religio Medici*'. He was also the author of *Hydrotaphia: or Urn Burial* and is said to have 'been celebrated among his contemporaries for *Vulgar Errors* (1646) in which he described many misapprehensions over natural phenomena and attributed some of these to

[3] *Buckley* v. *Rice Thomas* (1554) I Plowd. 118, 75 ER, KB 182.

[4] *Pickering* v. *Barkley* (1649) Style 132, 82 ER 587.

[5] *Buller* v. *Cripps* (1702) 6 Mod. 29, 89 ER 793.

[6] Ibid. See also *Chaurand* v. *Angerstein* (1791) Peake 61; *Kruger* v. *Wilcox* (1755) Amb. 252, 27 ER 168.

[7] *Fearon* v. *Bowers* (1753) 1 H. Bl. 364, 26 ER 214.

[8] *Ekins* v. *Macklish* (1755) Amb. 184, 27 ER 125.

[9] *R.* v. *Cullender* 6 How. St. Tr. 687. See also D. Thomas (ed.), *The Public Conscience*, vol. ii (Routledge, London, 1972), 4.

Satan's attempts to mislead the human race'.[10] Two elderly women of Edmundsbury, Rose Cullender and Amy Duny, were accused of bewitching some children who, thereupon, had violent fits. At other times they were struck deaf, blind, and dumb and 'at other times they would fall into swoonings, and upon recovery to their speech they would cough extremely, and bring up much phlegm, and with the same crooked pins and one time a two-penny nail with a very broad head, which pins (amounting to forty or more) together with the twopenny nail, were produced in court . . . Commonly at the end of every fit they would cast up a pin, and sometimes they would have four or five fits in one day.'[11] A Dr Jacob of Yarmouth had examined one of the children, having a reputation for helping bewitched children. A Dr Feavor examined another child and later gave evidence at the trial that he had no idea what caused the fits. Dr Browne's role seems to have been that of an expert observer, advising a court convened under the Lord Chief Baron, Mathew Hale, Mr Serjeant Kelygne (later Lord Chief Justice), Mr Serjeant Earl, and Mr Serjeant Bernard. It was only after the evidence had been given that he was asked to comment on the demeanour of the children and the accused: 'Asked for his opinion, he said he thought the fits might be natural, but that they were aggravated by the Devil in collusion with the witches; and that from his great knowledge he added that there was much vomitting of pins in Denmark.' He said that there had recently been a great discovery of witches in Denmark 'who used the very same way of afflicting persons'.[12] This opinion no doubt influenced the Lord Chief Baron in his conclusion 'that there were such creatures as witches he made no doubt at all'. Dr Browne explained that 'the children's fits and other afflictions had a scientific explanation. They were caused by the devil who did work upon the bodies of men and women, upon a natural foundation, (that is) to stir up and excite such humours

[10] Thomas, *The Public Conscience*, 62–3.

[11] Ibid. 68.

[12] 6 How. St. Tr. 687. See also T. R. Forbes, *Surgeons at the Bailey* (Yale University Press, New Haven, Conn. 1985), 26–7.

super-abounding in their bodies to a great excess.'[13] The women were convicted and executed on 17 March 1665.

Assessors and Advisers

The fact that the law in the books concentrates upon experts as witnesses obscures the fact that some courts carried on using experts as advisers sitting at the right hand of the judge well into the twentieth century. The two main courts which continued this tradition were the Admiralty and Patent Courts, where one or two assessors would sit alongside the judge. The Admiralty Court is usually portrayed as a highly idiosyncratic jurisdiction, a quaint backwater which has little or nothing of significance to contribute to our understanding of the law in modern society. However, it remains part of the structure of our legal system. Both the Admiralty and Patent Courts are widely used by the legal and business communities. They retain, in working order, the court-appointed expert model. This alone lends a lie to the commonly held view (1) that the court expert is a Continental invention; and (2) that the expert witness model is the only model operating in the English legal system. The truth of the matter is that the two models run alongside each other and have long done so. In fact, the existence of an alternative model within English law has provided a resource for both the supporters and detractors of the independent court expert model. In 1985, for example, the Attorney-General of Hong Kong voiced the idea that the lay jury should be abolished in complex fraud trials, and replaced by two expert adjudicators sitting alongside the judge. This idea was contrary to the rhetorical thrust of English law, which was to stop expert advisers trespassing on the judicial province.

In earlier times, experts were appointed by the court to sit alongside the judge and provide specialist advice. Specific reference can be found in the 1413 Patent Rolls[14] to the royal appointment of arbitrators to settle a dispute concerning prize money from ships. Twiss writes that on the *Inquisition of*

[13] 6 How. St. Tr. 687.   [14] 1 Hen. v. pt. 2, m. 14d.

Queenborrow in 1375, 'the circumstances and answers set out by the jury were laid down [and] the judges consulted the prud-homes of the merchants'.[15] Twiss suggests that this was the custom 'and has been used in all times'. He goes on to suggest that the practice adopted in the Admiralty Courts of using assessors was derived from the practice of foreign maritime courts. The office of assessor in Admiralty and Patent Courts fulfilled the same role as the court expert. In effect, only the terminology differed. McGuffie argues that assessors have a considerably earlier origin than the Charter of Henry VIII which incorporated Trinity House in 1514.[16] Twiss also refers to the sixteenth-century practice of judges of the High Court of Admiralty hearing cases with the assistance of two elder brethren of the Corporation of Trinity House of Deptford le Strand as assessors. Trinity House was the source of assessors for the Admiralty Courts. Roscoe's *Admiralty Practice* claims that the use of assessors 'rests upon ancient usage and not upon any statute'.[17]

In the fifteenth century a number of cases indicate that the Admiralty Court had adopted the practice of sitting with specialist advisers or assessors. In 1600, reference was made to Trinity House to ascertain whether damage to cargo was caused by weather or fault, and it was reported that 'there was no fault of marines'. In 1673, nautical experts were for the first time called in and concurred in a finding of Sir L. Jenkins in a collision case. Records indicate that in the eighteenth century assessors were asked by the court to decide upon such matters as whether a sentence ought to be sustained, altered, modified, or wholly reversed. Marsden notes judges summing up to the assessors and treating them as fellow adjudicators. An example of this practice can be found in the case of *Re Rumney and Wood*, where the president disqualified two pilots as being

[15] Sir T. Twiss, *The Black Book of the Admiralty 1871–76* (RS). Also the Records of the Admiralty Court of York; the Wreck Rolls of Leiston Abbey, 1377–1517.

[16] K. C. McGuffie, 'Notes on Nautical Assessors' (unpublished MS, Admiralty Registry, London).

[17] *Roscoe's Admiralty Practice* (5th edn. Hutchinsons, London, 1859). See also A. Dickie, 'The Province and Function of Assessors in English Courts', *Minn. Law Review*, 33 (1970), 494, 499; *The Queen Mary* (1948) 80 Ll. L. Rep. 609, 612.

'unworthy, unfit, unskilled, inexperienced, lazy, negligent and careless men'. In 1601 there is another example in the case of *Leighton* v. *Peter Moore*, where it was the judgment of the assessors that Moore was a 'skilful mariner and at no fault'; a further example may be found in *Cornwallis* v. *Noden* (1673).[18]

In essence, the court expert acted in much the same way as the special jury. Both were relatively free from the restraints of judicial control. They could not be cross-examined, their advice was given in private, it was not usually disclosed to the parties, nor were judges under any compulsion to make a written note of it. In addition, it has been the practice in England not to allow expert evidence to be brought on the behalf of the parties where the court is assisted by assessors.[19]

The use of assessors began to die out in the United States in the nineteenth century.[20] In the United Kingdom, the demise of summing up to the assessors in the Admiralty Court and the refusal to allow traders to comment upon a trade restriction in 1899[21] were symptomatic of judicial concerns about the proper province of experts. The fact was that judges were summing up to assessors, treating them as fellow judges and bowing to their judgement. This was beginning to prove an embarrassment. Judges also made the mistake of admitting that they were unable to make decisions on their own; they felt compelled to follow the advice of their assessors. Experts were taking over the role of the judiciary. This fear reached its height during the nineteenth-century push towards a more rational legal system. A statute of Victoria[22] specifically provided that the House of Lords could require the assistance of nautical assessors when hearing Admiralty matters. This enactment was no more than a recognition of the fact that the advice of such specialists was already heavily relied upon by the Admiralty

[18] *Cornwallis* v. *Noden* (1673) cited in R. G. Marsden, *Select Pleas of the Admiralty* (SS), vol. i: *1390–1404*; vol. ii: *1527–1545*; vol. iii: *1547–1602*; See also R. G. Marsden, *Law and Custom of the Sea (1205–1648)*, 2 vols. (Navy Records Society, MDCCCLXV: 1915, 1916).

[19] See *The Ann and Mary* (1843) 2 Wm. & Rob. 189, 166 ER 725; *The St. Chad* (1965) 1 Ll. Rep. 107; *The Fritz Thyssen* (1967) Ll. Rep. 104.

[20] *The City of Washington* (1875) 92 US 31.

[21] *Haynes* v. *Doman* (1899) WN 65; (1899) 1 Ch. 67, per Lindley, J.

[22] 54, 55 Vict.

judiciary. As the nineteenth century wore on it became increasingly clear that the balance of responsibility for decision-making was shifting towards the assessor. It was to counter this idea that in 1850 Dr Lushington felt impelled to draw the line between the role of the assessor and the role of the judge:

I have never yet pronounced a single decree when I was assisted by Trinity Masters, in which I was not perfectly convinced that the advice they gave me was correct and if I had entertained a contrary opinion, notwithstanding all their nautical skill and experience, I am clearly of the opinion, having deliberated much upon the question, that it would be my duty to pronounce such an opinion. The court never allows persons, perhaps inexperienced in the administration of justice, to raise inferences from a supposed state of facts. It fortunately has happened that in but very few instances, there has been a difference of opinion between myself and the Trinity masters, and in no case whatever have I pronounced any judgment except it was my own.[23]

In *The Beryl* (1884) Brett also felt moved to underline the subsidiary role of the assessor:

The tribunal which has to try the case is the judge himself and the judgment is his and his alone. The assessors who can assist the judge take no part in the judgment whatever; they are not responsible for it and have nothing to do with it. They are there for the purpose of assisting the judge by answering any question of nautical skill. The judge is bound to give great weight to the opinion of assessors but at the same time if he does not think their view is right, he is not bound to follow it . . . [But] it would be impertinent in a judge not to consider as almost binding upon him the opinion of the nautical gentlemen who, having ten times his own skill, are called in to assist him.[24]

This view nicely summarizes the dilemma of judges involved in hearing Admiralty matters. Scrutton, LJ, had stated in several cases that he felt bound to follow the advice of his assessors. This earned him a rebuke from the House of Lords which reminded him that 'the assessor, being nothing but an

[23] *The Ann and Mary* (1843); see also *The St. Chad* (1965).
[24] *The Beryl* (1884) 9 PD 4, 137, per Brett, J.

adviser, it could not be right for a judge to surrender his opinion'. In a further case, the higher court said that Scrutton 'appears to have surrendered his own view to that of his nautical assessors, and it is clear that it is the duty of the judge to form his own judgment whatever that judgment may be'.[25] In *The Tovarich* (1930) Scrutton, LJ, again drew attention to the judiciary's dependence upon expert advice. Commenting on the practice of using a different set of assessors at every level of appeal, he spoke of the assessor as a very peculiar sort of witness:

> The judge in the Admiralty talks to them and gets information from them. The parties do not know what the witnesses are telling the judge; they have no opportunity of cross-examining the so-called witnesses. Indeed, in the Admiralty Court, the practice is not followed which we—in obedience to the direction of the House of Lords follow—the practice of asking questions in writing and obtaining answers in writing, and sending them up to the superior court. One starts, therefore, with two witnesses whose evidence the parties do not hear, and whom the parties have no opportunity of cross-examining, and the case then comes to this court, and we have to decide the case with two more witnesses whom the judge below did not hear.[26]

Assessors were in fact being used as judges at the very highest levels of the legal process. Scrutton, LJ, made his point in a more ironic tone when commenting in *The Llanelly* that 'The art of seamanship and navigation appears to be a very exact science. I only say that because this is one of those numerous cases in which the two experienced gentlemen who advise us have come to exactly the opposite conclusion. In cases which have gone to the House of Lords, a third conclusion has sometimes been provided by the experienced men who advise their Lordships' House.'[27] In other words, nautical science must be very complex if it required the expertise of so many men of science to unravel its mysteries. There was an increasing tendency to rely upon appeals from assessor to assessor rather than from judge to judge. The practice had become so com-

[25] *The Llanelly* (1926) Ll. L. Rep. 37.
[26] *The Tovarich* (1931) AC 121.
[27] *The Llanelly* (1926).

mon that in time the House of Lords felt compelled to correct this tendency. It went out of its way to bring home to the lower courts that it would be 'intolerable if appeals were treated as being not from one judge to another but from one assessor to another'.[28] The judiciary was not to allow itself to be bypassed by such practices. It should take firm steps to adhere to the proper relationship between the judiciary and its expert advisers:

Speaking for myself, I come to the same conclusion as the majority of assessors here and I should have done so unassisted by the opinion of any assessor. But as it stands the case raises in an acute form the propriety of what has been called an appeal from assessors to assessors. There is no hierarchy of assessors. They occupy much the same position as do skilled witnesses with a difference, that they are not brought forward as the partisans of one side or the other. If assessors differ, the court must make its own choice. In every case the responsibility is with the court. Speaking for myself, I shall always ask the assessor as little as possible as it is much oftener a question of common sense than a question of seamanship. That the different assessors are at variance is much more of a hindrance than an assistance.[29]

Supporting Lord Dunedin in *Australia* v. *Nautilus*, Lord Sumner reiterated the judiciary's responsibility not to surrender its province to the assessor. Any uncertainty about the relationship between the two players in the legal process was dispelled:

[the judge] might be tempted in his perplexity to renounce the task of judgment. This, however, he must not do. Technical advisers are not judges even of such issues as these. The appointed court must exercise its function of deciding, and find consolation in a consciousness any rate of blank impartiality. An opposite road of escape might seem open—namely, that of accepting, as of course, the advice of assessors consulted. In reality, there is no escape here. This is what leads to the 'intolerable situation' as it is called of 'appeals from assessors to assessors'. The phrase, pungent as it may be, requires

[28] *Owners of the SS Australia* v. *Owners of the SS Nautilus* (1927) AC 145, 150, per Lord Dunedin.
[29] Ibid.

some explanation. Strictly it is a figure of speech, for there is no such thing. It is really a criticism of the conduct of one court, or of both courts, when the court adopts a certain attitude towards its assessors. That attitude is the court's surrender to the assessors of the judicial function of itself deciding the issue, however technical it may be. Authority for the proposition that assessors only give advice and that judges need not take it, but must in any case settle the decision and bear the responsibility, is both copious and old. They are entitled and even bound, though at a risk of seeming presumptuous, to give effect to their own view. Such being the position of judges, what is that of the assessors? They are technical advisers, sources of evidence as to facts. Now, howevermuch Admiralty judges may from time to time seem to have treated the Elder Brethren as members of the tribunal, however informal these consultations may have been, the principles stated above have never been in doubt. The function of the Court of Appeal is to urge upon the courts below and to impose upon itself the duty of making up its own mind alike on questions of nautical skill and on the value of the advice given.[30]

Australia v. *Nautilus* was taken as a powerful reinstatement of the judicial province. Judges were advised to steer a careful and correct course through difficult waters, but since they lacked any practical advice on how to do this, the disciplinary effect of *Australia* v. *Nautilus* was short-lived. The law in practice imposed its own *realpolitik*. By the 1960s, the judiciary again found itself in something of a cleft stick. It did not wish to be seen to be bound by the advice of its experts. On the other hand, the legitimacy of its decisions within a specialist client community depended upon their being backed by the appropriate expertise. In a complete reverse of its former position, the judiciary was now castigated for failing to place a proper value on the advice of assessors. Thus in *The Marinegra* (1960) Wilmer, LJ, was criticized for not having given good enough reasons for going against the advice of his assessors.[31] He himself reiterated that a judge ought not to disregard the opinion of his assessor. In the case of *The Magna Charta* in

[30] *Owners of the SS Australia* v. *Owners of the SS Nautilus* (1927) AC 145, 150, per Lord Dunedin.

[31] *The Marinegra* (1958) 2 Ll. L. Rep. 385; CA (1959) 2 Ll. Rep. 65; HL (1960) 2 Ll. Rep.

1871, the Court of Appeal had specifically stated that the opinions of assessors should receive great weight, adding, almost as an afterthought, that 'it is, however, the duty of the court to decide the case. The judge is bound in duty to exercise his own judgment, and it would be an abandonment of his duty if he delegated that duty to the person who assisted him.'[32] In 1971 the emphasis swung in the other direction as the House of Lords castigated the Court of Appeal for its uncritical acceptance of the advice given by assessors in the court of first instance.[33] This was said to be a failure by the court correctly to exercise its own judgment on the whole circumstance.

The needs of commerce made heavy demands in particular upon the Admiralty and Patent Courts. With their old practice of sitting with assessors still intact, they were well placed to settle the legal disputes between businessmen. Yet they continually faced uncomfortable questions about their expertise, and in some quarters the idea of specialist tribunals to replace the courts found favour. Alternatives of this kind rapidly grew in popularity in the twentieth century, as the competence of these courts to handle rapid and complex scientific advances was put to the test and found wanting. Freckleton cites a case from 1935 when the court spoke enthusiastically about the use of assessors: '[if] full justice is to be done in the adjudication of patents, the judges should have associated with them in a confidential and intimate capacity unbiassed, thoroughly competent, scientific aides. It is becoming more and more apparent that the courts as now constituted can rarely reach just conclusions in matters where new and complicated scientific truths must be interpreted and serve as the only guide posts.'[34] However, despite the provision by section 88(1) of the Patents Act 1952, as well as under Order 103, Rule 27, of the Supreme Court Rules, Freckleton notes that relatively few assessors have been appointed to the Patent Courts. In 1973

[32] *The Magna Charta* (CA) (1871).

[33] *The Statue of Liberty* (HL) (1971) 14 Asp. MLC 321; (1971) 2 Ll. Rep. 277.

[34] *Adhesives Pty. Ltd.* v. *Aktieselskabet Dansk Gaeringindustri* (1936) 55 CLR 523, 580, per Rich, J.

the Court of Appeal appointed a scientific adviser for the first time since 1935, without first gaining the consent of the parties.[35]

This juggling act between the needs to protect the province of the judiciary and at the same time please a specialized client community was as evident in the Patent Courts as it was in Admiralty. Judges felt increasingly beholden to their expert advisers for advice on scientific matters. Patent issues posed hard problems. They arose in complex areas at the forefront of scientific advances. Their very novelty was an issue for the court to decide. As Freckleton points out, it is not overstating the situation to say that a welter of new scientific techniques and theories began to besiege the courts.[36] How was the judge in the Patent Court to assess scientific claims which were novel or contentious? The means of verifying scientific claims adopted by the courts had traditionally focused either upon the trustworthiness of known experts, or upon the general acceptability of a scientific theory within the scientific community. It was partly because courts of law assumed that consensus could be found that they persisted in pursuing agreement even where genuine disagreement existed.

The same kind of thinking may still be seen in legal and quasi-legal proceedings in which the status of certain scientific theories is controversial. Planning inquiries such as the Windscale and Sizewell B Inquiries demonstrate the inability of the judicial frame of mind to accept that men of science can quite normally disagree not only about how to interpret the facts but also on the facts themselves. Using a fellow scientific expert as assessor in such inquiries has not resolved this issue, and where agreement is not forthcoming within a given time-frame, one alternative has been to impose it. Which view prevails depends upon a number of things, only one of which is whether it fits in with existing paradigms. Whether or not a view will succeed is dependent upon a number of cultural fac-

[35] See *Valensi* v. *British Radio Corporation* (1973) 52 RPC 337. See also *Mullards Radio Valve Co. Ltd.* v. *Philco Radio and Television Co. of Great Britain* (1969) 52 RPC 270 CA; *Re Nossen's Patent* (1969) 1 WLR 638; (1969) 4 FSR 403; (1969) All ER 775.

[36] I. R. Freckleton, *The Trial of the Expert* (Oxford University Press, Melbourne, 1987), 165.

tors as well as upon the internal structure of debate within the scientific community. In the case of novel patents, no such community of consensus was available to the Patent Courts. Consensus is not, in any event, a reliable guide to the soundness of scientific knowledge, and the scientific community is well known for its conservative resistance to novel developments. Major commercial enterprises ensure that considerable material resources will be spent out-proving rival claims. Parties with sufficient material resources were (and are) better able to muster support. For cases turning on the status of scientific developments, experts have proved rather unhelpful. Where support for a particular point of view comes it is usually challenged by equally eminent experts, and this only serves to exacerbate the conflict.

Looking at disputes in the patenting of radio and television, it becomes clear that there was no neutral expert available to help the court. Most of the experts who were competent in the field were employed by firms with direct interests in developing their own patents. They could not, therefore, be thought neutral. It was from this small circle of competitors that the courts were forced to recruit their expert advisers. In such circumstances, finding an independent assessor proved virtually impossible. Despite their limited competence, Patent Court judges thus became the ultimate arbiters of the commercial application of science. It is difficult to imagine a more powerful role and one which expresses so vividly the domination of law over science. Judges unversed in advanced technology had the power of life and death over intellectual property and its commercial application. Certainly some patent actions involved the application of highly technical detail and advanced scientific theory. Though it might fall to the scientific community to judge whether these theories were internally consistent and coherent, it was the law which held the final power over their dissemination in practical form.

In this situation the judiciary reached the limits of its competence. Inexpert Patent Court judges were required to make all sorts of complex judgments about the status of theoretical and practical developments in fast-changing fields of science and technology. Large amounts of money and rights to manufacture and distribute products depended on the outcome of disputes. There was a sense of urgency which demanded that these issues be decided quickly. Given this minefield of scientific and commercial factors, and the struggle to apply legal principles of intellectual property at the frontiers of knowledge, it is hardly surprising that judges in the Patent Courts felt inadequate. Their competence and credibility faced constant challenge. Like the judges in Admiralty, they were often tempted to adopt the position of their experts as their own, and throughout the 1930s, 1940s, and 1950s, Patent Court judges echoed the dilemma faced by their Admiralty Court colleagues.

In the commercial courts, one remedial course of action was to bolster the expertise of the judges. Some judges went to great lengths to increase their competence. Other judges acquired lengthy experience at the specialist Bar. To some degree these efforts paid off. The movement of business away from the courts in favour of specialist arbitration found favour for a time, but it soon became evident that expert arbitrators were in short supply. Their services were booked up months in advance. Facing delays at arbitration, commercial clients slowly began to take some of their business back to the courts. Moreover, courts of law maintained their formal and structural superiority, since they alone had the power to enforce decisions. This was a power which arbitration lacked.

Taken together, the experience of the Admiralty and Patent Courts reveals a judiciary highly dependent upon scientific expertise. How did the judiciary cope with the prospect of appeals going not from judge to judge but from assessor to assessor? How did it cope with having the limits of its competence revealed? Having given experts *de facto* judicial status the judiciary followed this up with efforts to put them back in their place. This at least kept up appearances, even if in reality the judiciary continued to defer to the opinions of experts. However, just as the English judiciary was trying to keep experts out of the judicial province, judges in the United States were seeking to elevate experts to precisely that position.

Court Experts: The American Route

In a sense, what the American courts did was to reinvent the wheel. By the turn of the century, American jurists, discontented with the so-called battle of the experts, were considering the merits of using court experts. Analysis of the debate in legal journals reveals that they saw the court expert as a radically new model for the reception of expert evidence. Court experts enjoyed an immense but brief popularity in the United States at a time between 1880 and 1920 when America was generally more receptive to all things scientific. This included scientific management, scientific housekeeping, scientific public administration, and scientific social work. America saw itself as a nation of science in which scientists could be called upon to help solve the nation's problems. This extended to the courts, where scientists were seen as particularly suited to mediate between conflicting parties. In 1893 the president of MIT had described scientists as men of sincerity, simplicity, fidelity, and generosity of character, noble in their aims and in their efforts. They were seen as totally objective, altruistic, and impartial.

It was in this climate that American law reformers rediscovered the court expert. By the 1940s and 1950s, a considerable body of case law had been established which allowed judges to appoint their own court expert.[37] New York established an Impartial Medical Expert Plan to deal with personal injury cases. Scientific courts, in which scientists and not judges adjudicated the technical issues, were recommended as an alternative procedure for resolving complex cases.[38] The Federal Rules of Evidence gave effect to the Model Expert Testimony Act, and in Rule 706 it specified the rules for appointment of a court expert by the court, without the consent of the parties. The New York Impartial Medical Testimony plan was endorsed by the American Bar Association and a comprehensive scheme for court-appointed experts was initiated with the

[37] See e.g. *Polulich* v. *Schmidt Tool Die & Stamping Co.* (1957), per Gaulkin, JCC, 29–36.

[38] D. Sherman *et al.*, 'The Use of Expert Witnesses in American Courts' (draft MS; 1990). Also 'Procedures for Decision-Making Under Conditions of Uncertainty: The Science Court Model', *Har. J. Leg.* 16 (1979), 443.

adoption of Local Rule 22 of the Federal Rules of Criminal Procedure 1946. In America in the 1930s the universality, communality, egalitarianism, and rationality espoused by proponents of the scientific enterprise were juxtaposed against the irrationality of authoritarianism prevalent in some parts of Europe. Mulkay has noted how scientists draw freely on this rhetoric, whereby they emerge as the 'ideal diplomats'.[39] It was in this context of scientists as philosopher kings bringing about the end of ideology that the idea of the court expert took hold. American law reformers were generally enthusiastic about incorporating science into the judicial process. The idea of government by experts found wide appeal and the image of science prospered.

Problems arose, however, when it became clear that even when they were co-operating rather than competing medical men differed categorically in their opinions. It was discovered that scientists disagreed amongst themselves even where they were not paid hirelings. Moreover, they disagreed not only about their interpretation of the facts but also about the facts themselves. Thus, for example, they argued over whether what was shown on an X-ray really was a fracture. To the law, it seemed strange that experts could not even agree on such basic things. They prolonged the dispute instead of curtailing it.

The popularity of the court expert model was relatively short-lived.[40] The main objection was that it was anti-democratic. There was also no guarantee that the court expert himself represented the impartial and accepted view of the profession as a whole. Court experts, just as much as expert witnesses, were likely to be partial in their exposition of particular schools of thought. Moreover, the authority of the court expert's view was great. The jury might well regard such an opinion as objective and impartial when in fact this would be an inaccurate impression.[41] Such weighty opinions, sanctioned by the court, would acquire an aura of infallibility which effectively undercut the jury's function. The American legal profession spent some time during the 1940s and 1950s agonizing

[39] M. Mulkay, *Science and the Sociology of Knowledge* (Allan & Unwin, London, 1979).

[40] Sherman et al., 'The Use of Expert Witnesses'.

[41] Ibid.

over the court expert model and looking to the Continent for encouragement. It found only other kinds of objections and dissatisfactions. The new model, it seemed, was as beset by problems as the old one. Gradually, American courts reverted to the use of experts as witnesses. This was seen as an attempt to meet the ideals of a full-blown due process and rule of law model. American courts did not turn their backs on experts nor did they place them in a position of supreme power. They adopted a more flexible—but also more rigorously policed—approach to the reception of expert evidence within the confines of the trial. They opted for a model which surrounded expert evidence with a multitude of procedural requirements, and which demanded a vigorous contesting of the status and credit of the expert. This was a model based on scepticism about expert claims. Unlike the English model, it did not expect impartiality on the part of expert witneses.

English Resistance

Why was there such a different attitude towards scientific experts in these two common-law jurisdictions? The English reaction to the revived notion of court experts was altogether more cautious. The argument that it should adopt the more rational Continental system of court experts found little favour with the English judiciary. The idea of making experts more powerful by placing them alongside the judge was not welcome. The English legal system had spent several hundred years moving experts away from that position. It had put a good deal of effort into working out a logic which would give the judiciary control over the verdict by separating the juror and witness roles. It had coped with the internal logic of this separation and the impracticality of the fact/opinion rule. It had come to terms with the need to bend the self-same rule in order to admit the evidence of expert witnesses. And at the end of all this it was still fighting a defensive action against expert incursion into the judicial province.

The English judiciary made much of the idea that independent court experts were a foreign invention. The court expert model stemmed, it argued, from the practice of Continental courts, especially in France and Italy, where experts sat alongside judges. Things were different in these countries. The court expert was entitled to make investigations pre-trial, and might be called to assist the judge during the trial itself. Experts in Continental countries were selected from the established academic or professional institutions, some of which had a longstanding obligation to furnish the courts with court experts. There was, it was said, no such model in the United Kingdom (though Trinity House had supplied assessors to the Admiralty Courts for hundreds of years). It was further argued that experts selected to assist the courts might not provide a fair and representative view of the issues. They might lack competence and integrity; they might favour a particular theory which could prejudice the assessment of the case; being court appointed might create the misleading impression of unquestionable impartiality.[42] Such a system was said to have no place in an English tradition which had long resisted trial by expert. Long experience was said to have demonstrated that trial by jury was best fitted to the English character. Such a radical change must be treated with utmost caution. Law in England had developed over hundreds of years. It was not to be tampered with lightly. To introduce foreign notions such as independent court experts would be to strike at the very roots of English justice. This was a remarkable position for a system which had known and used court experts for almost 500 years.

In both the United States and English jurisdictions, a compromise proposal was mooted which preserved the status quo whilst appeasing those who called for specialist courts. This compromise allowed the parties, rather than the court, to agree between them on the appointment of a court expert. They could (1) scrutinize his evidence by way of cross-examination; and (2) introduce the evidence of their own experts to refute it. In the United States, Federal Rule 706 permits the parties to participate in the selection of a court-appointed expert and to make nominations. Where the parties

[42] J. Basten, 'The Court Expert in Civil Trials: A Comparative Appraisal', *Minn. Law Review*, 10 (1977).

do not agree, the court may appoint its own candidate. Whether or not to disclose to the jury that an expert is court appointed is a matter for the discretion of the judge. In England, the judge in civil actions has the power to appoint a court expert in any non-jury case, on the application of the parties. The function of such an expert is said to be to inquire and report upon any question of fact or opinion, where a question for an expert arises.[43] There is no specific provision for the appointment of a court expert in criminal proceedings but a judge may call a witness not called by either the prosecution or the defence, without their consent.[44]

Why has this compromise position not been more widely adopted? Part of the explanation is the unsatisfactory nature of its results for those seeking a more rational and expeditious means of resolving disputes in specialist areas. It is said to have a number of drawbacks. Instead of reducing the number of experts in court the compromise model increases them. It has thus found little favour with the judiciary already afraid of trial by expert. Instead of reducing the potential for conflict amongst men of science it in fact increases it. The compromise model thus fell into disfavour with the parties to the case, who expected to achieve consensus. Lord Denning drew attention to this problem in 1962:

> It is said to be a rare thing for it to be done [appointment of a court expert]. I suppose that litigants realise that the court would attach great weight to the report of a court expert, and are reluctant thus to leave the decision of the case so much in his hands. If his report is against one side, that side will wish to call its own expert to contradict him, and then the other side will wish to call one too. So it would only mean that the parties would call their own experts as well. In the circumstances, the parties usually prefer to have the judge decide on the evidence of experts on either side, without resort to a court expert.[45]

[43] P. Murphy, *A Practical Approach to Evidence* (Blackstone, London, 1980), 299; Rules of the Supreme Court, o. 40 r. 1; also rr. 1(2) and 1 (3); o. 38 r. 16; Supreme Court Act 1985, s. 70; o. 40 r. 15; see also Patent Act 1977 and RSC o. 33 (*b*) and o. 103 (*b*); Annual Practice RSC o. 103 r. 27.

[44] *R.* v. *Liddle* (1928) 21 Cr. App. Rep. 3; *R.* v. *McMahon* (1953) Cr. App. R. 95.

[45] *Re Saxton Deceased (Johnson and Another* v. *Saxton and Another)* (1962) WLR 968, 972.

Other obstacles have also been cited as causes of the unpopularity of the court expert. Experts chosen to sit as court experts are said to dislike being drawn into public controversy with colleagues. There is no compulsion upon anyone to use a court-appointed expert. The initiative to use (and pay for) one has to come, not from the court, but from the parties themselves. The threat of excessive judicial respect for the court expert deters the parties from initiating such a request. The failure of the compromise position is, then, partly due to inertia and partly to a distinct lack of judicial will to make it work. In any event, the attraction of the court expert model for the English judiciary lay in its promise to bolster the flagging competence of the judiciary. It lost its attraction when it became clear that one side-effect of this was the displacement of the judiciary by a host of experts in court, with the prospect of appeals going from expert to expert rather than judge to judge.

Interest in the court expert model thus flourished for a brief time at the turn of the century and lasted for another forty years or so in the United States. It grew out of the law's struggle with expert witnesses and, given a climate which favoured administration by experts, it seemed to offer a modern solution to all the law's problems. Prior to the nineteenth century, experts had posed relatively few problems—there were fewer of them and they tended to be drawn from a select circle. Science being fairly new, it was quite acceptable for judges to enter into arguments with experts about the adequacy of scientific knowledge. Science and technology had yet to become a fragmented and mysterious set of specialisms which denied the layperson access to their inner workings. To extend a point raised by Wynne, when science and technology go beyond the stage where they at least seem to be within the public ken to the point where neither knowledge nor control is any longer a real possibility, then one is faced with either a total acceptance of scientific and technological developments, or a total rejection of them. There is no in-between; there is no informed public debate.

Judges stood at the crossroads between these two polar positions. The vast expansion of science and technology meant the appearance of new knowledges and more experts. Court assessors could still be drawn from a select band of people; their appointment was still within the behest of the judiciary. Expert witnesses were recruited by the parties from a wider pool. In the United States, the vast size of this pool of experts and the relative absence of a scientific élite made it impossible for courts to draw experts from within their own knowledge. Their controls had therefore to differ from those imposed by the English courts, which could still draw on a relatively circumscribed circle of experts. To a public exposed to a multitude of bogus expert claims, expert status was something which had to be tested and earned before cynicism could be overcome. American lawyers continually contest an expert's qualifications, his standing in the expert community, and the validity of his answers, especially where these are answers to complex hypothetical questions. Expert status must be achieved. In England, by contrast, expert status was ascribed. This approach was more in keeping with the existence of an established legal élite.

The refusal of the English legal profession to engage seriously with the claims of experts in the first half of the twentieth century left an unfortunate legacy. American courts, having taken the claims of scientists seriously, tried using experts as judges and found them wanting. As a result of this experience, American courts adopted a more sceptical view of expert evidence. In England, the courts never tested the claims of experts in this way, with the result that English courts never developed the more sceptical view of scientistic claims. In the absence of such informed scepticism, it was open to English courts to continue with their highly unrealistic expectations of experts. Moreover, the arguments against court experts were seldom aired in the very courts which were highly dependent upon court experts. They needed the participation of experts if they were to reach credible decisions and retain their client communities. The judiciary in this situation could not afford to be overly critical of its expert advisers. The need to protect judicial hegemony was reflected in a rhetoric which played up the symbolic nature of the judicial role, but which neither imposed nor required active compliance at the level of practice. Even where, as in the Admiralty Courts, the practice of judges surrendering to assessors became highly visible, the result of the rhetorical admonition seems simply to have been a judiciary not any less likely to surrender its duty but more circumspect when doing so.

4 The Challenge of Science

With the creation of the new model jury and the new witness role came expert witnesses. Legal texts cite the case of *Folkes* v. *Chadd* as the precedent for the acceptance of expert testimony on opinion in English courts despite the fact that experts had been giving opinion evidence for almost a century beforehand. In 1678, for example, 'some of the most eminent physicians in England' appeared as defence expert witnesses for Spencer Cowper.[1] Whilst 'the prosecution's medical witnesses were virtually unknown outside their home counties, . . . Cowper's physicians and surgeons were famous throughout the kingdom'.[2] Forbes's study of the Old Bailey records for the period tells us that these experts included the physician to Christ's Hospital, Sir Hans Sloane, who was later to become president of the Royal Society, Sir Samuel Garth, 'successful physician and literary and political figure', a Dr Crell, and the eminent anatomist and surgeon William Cowper.[3] A Scottish case from 1750 also appears to have involved expert pathologists as witnesses, supplying the court with a post-mortem report in a murder case.[4]

[1] See A. Rosenberg, 'The Sarah Stout Murder Trial: An Early Example of the Doctor as Medical Expert Witness', in C. R. Burns (ed.), *Legacies in Law and Medicine* (Science History Publications, New York, 1977), 230.

[2] Ibid.

[3] T. R. Forbes, *Surgeons at the Bailey* (Yale University Press, New Haven, Conn., 1985), 47–8.

[4] The following report of the case is taken from the Scottish Records Office JP 12/2/1: 'Doctor Patrick Maxwell at Burnhead and James and John Mowats surgeons at Langholm went to view and inspect the corpse of a man lying at Stadingstean at Woodhouselees ground, who blooded to death there, and was thought to come by that hasty death by some violence or hurt which Lancelot Brown and James Brown, his father in the same James Brown's house. They did find upon opening the Chist that the lungs did strongly adhere on the right to the pleura, but little or no corruption, and no blood to be found in them, and but little in the largest blood vessels; but could discover no appearance of violence, or vessels broke, or extravast blood in the cavity of the thorax, nor any corrupt smell; but when they came to the stomach, found blood to the quantity of a pound or more, and but little corrupted, but how such quantity of blood came there, was impossible to find out, both from his having been dead three days and the intestines tending much to putrefaction; but if subject to any preceding

The substance of *Folkes* v. *Chadd* concerned an action for trespass in which the issue was the cause of a harbour filling up.[5] The plaintiff produced an engineer, Mr Smeaton, author of the Eddystone Lighthouse. His evidence was well received, but caused problems for the opposing party, who objected to it on the grounds of its being 'a matter of opinion what had caused the silting up of the harbour', whereas the jury should base its verdict only upon facts. Lord Mansfield was of a different persuasion:

It is objected that Mr Smeaton is going to speak, not to the facts, but as to opinion. That opinion is, however, deduced from facts which are not disputed—the situation of banks, the course of tides, and of winds, and the shifting of sands. His opinion, deduced from all these facts, is, that, mathematically speaking, the bank may contribute to the mischief but not sensibly. Mr Smeaton understands the construction of harbours, the causes of their destruction, and how remedied. In matters of science no other witnesses can be called . . . the cause of the decay of the harbour is a matter of science, and still more so, whether the removal of the bank can be beneficial. Of this, such men as Mr Smeaton alone can judge. Therefore we are of the opinion that his judgment formed on facts, was very proper evidence.[6]

In a sense, *Folkes* v. *Chadd* was simply a formal admission of what was in fact already the day-to-day practice. The presiding judge, Lord Mansfield, cited several previous occasions from within his own knowledge on which experts had testified to opinion. With regard to the navigation of ships, for example, he stated that:

The question then depends on the evidence of those who understand such matters; and when such questions come before me, I always send for some brethren of the Trinity House. I cannot believe that when the question is whether a defect arises from a natural or artificial cause, the opinions of men of science are not to be received. Handwriting is proved everyday by opinion . . . Many nice questions

blooding, which was reported, they did think that by the scuffle and struggling with the heat of passion, might have put the blood in a more violent motion than ordinary, and by that enlarged whatever ruptured vessels might be and so poured in these contents as they had found it in the stomach, but in every other case quite healthy and lusty.'

[5] *Folkes* v. *Chadd* (1782) 3 Doug. 157, 99 ER 58.
[6] Ibid.

may arise as to forgery and as to the impression of seals. In such cases I cannot say that the opinion of seal makers is not to be taken.[7]

Folkes v. *Chadd* simply legitimized what was clearly already a long-standing practice. It eased the law's difficulty by making expert evidence the exception to the rule excluding opinion evidence. There was no real philosophical justification for doing this. There were, however, compelling practical reasons. According to Thayer, the law found experts necessary 'because, being men of skill, they can interpret phenomena which other men cannot, or cannot safely, interpret. They "judge" the phenomena, the appearances or facts which are presented to them, and testify to that which in truth these signify or really are; they estimate qualities and values. We say that they are testifying to opinion. In truth they are "judging" something and testifying to their conclusion upon a matter of fact.'[8]

Lord Mansfield's judgment was translated into the general legal precept that the opinions of scientific men upon proven facts may be given by men of science within their own science. The importance of *Folkes* v. *Chadd* lies in this legitimization of the expert's status as a special sort of witness. But this new status was not all good news for experts. Whereas their opinions once had judicial status, they were now 'a mere witness's declaration'[9] to which the judge applied the 'appropriate legal rule'.[10] Moreover, as a juror and a court expert, the expert had enjoyed relative independence. As a witness, he fell squarely within the ambit of judicial control. Even though experts had been elevated to special witness status they had nevertheless been demoted in terms of their power to influence trial outcomes. *Folkes* v. *Chadd* underlined the fact that the place of the new science was in the witness-box, as an aid to justice rather than its dispenser. In one and the same move, the law underlined the structural diminution in the expert's role whilst hailing the usefulness of experts and their right to a

[7] Ibid.

[8] Thayer, *Treatise on Evidence.*

[9] J. B. Thayer, *Treatise on Evidence at Common Law* (Rothman Reprints, New York, 1969), 196.

[10] Ibid.

special place in legal proceedings. The degree of rhetoric used to celebrate the status of experts in *Folkes* v. *Chadd* stands in inverse relationship to the power experts wielded.

Post-Folkes *v.* Chadd

In the period immediately following *Folkes* v. *Chadd* the new sciences expanded. They proved increasingly versatile and useful to the point where a belief in science virtually became a national ideology: 'scientific methods and standards could achieve anything to which the Western World had been the heir ever since; it brought with it the now-familiar cult of expertise and professionalism which affected even government to an extent previously unparalleled.'[11]. *Folkes* v. *Chadd* arose at a time when the scientific and industrial revolutions were beginning to produce important results. It was a period of economic growth and political instability in post-Restoration England:

> There was clearly an indirect link between science and economic life, for the prevalent innovative attitudes in science were mirrored in the fertility of invention shown by patent applications and in a widespread optimism about the potential for improvement. Indeed, there were hopes in scientific circles that the link might become more direct, with national prosperity and useful, scientific knowledge advancing hand in hand. Thus it might be possible 'to render England the Glory of the Western World, by making it the Seat of the best knowledge as well as it may be the Seat of the greatest Trade', with merchants patronising and assisting intellectual activity, whilst it was also hoped that the findings of the new science could improve techniques in agriculture and industry.[12]

The new science found favour amongst London's 'culture of virtuosi'.[13] This provided a receptive and significant audience for scientific matters. Puritanism had laid clear the possibility of transforming nature. Indeed, the transformation of nature as

[11] M. Hunter, *Science and Society in Restoration England* (Cambridge University Press, Cambridge, 1981). See also M. Purver, *The Royal Society: Concepts and Creation* (Routledge, London, 1967).

[12] Hunter, *Science and Society*, 192.

[13] Ibid. 5.

a means to the material and spiritual improvement of mankind became a virtue and a duty. Several writers have linked the value-orientation of seventeenth-century society more generally with the rise of capitalism and technology in the West, and, as we have seen, Harding identifies this moment as one of particular significance in the rise and demise of the social agenda of science. At the same time as the judiciary was seeking to establish control over the realm of legal work, Newton was writing his *Philosophiae naturalis principia mathematica* and *Opticks*, and Bacon had published major works in the 1620s, principally *New Organon* and *The Great Instauration*. Harvey's *Anatomical Exercitations* was published in 1653 and in the 1660s Charles II set the royal seal of approval on the foundation of the Royal Society. Newton's account of the universe was a direct threat to the established order. It presented the universe rather like the intricate workings of a huge mechanical clock. As Harding points out, a good deal of information about the world was also being generated by skilled artisans who fell outside the circle of gentlemen scientists. The two met, however, in the coffee-houses, where 'scientific enthusiasts' met. It was a commonplace that these were patronized by people of diverse social backgrounds: 'Gentry, tradesmen, all are welcome hither and may without affront sit down together.'[14] The impetus to develop ways of manipulating nature seems to have come less from aristocratic circles than from the new bourgeoisie. The Royal Society was something like a gentlemen's club, a social gathering of people 'vaguely interested in science but not actively concerned with it'.[15] But its search for large financial endowments led it to stress the utilitarian aspect of scientific endeavours in its appeals to the landed gentry.[16] Lack of response, however, led the Royal Society to become what it had already been from the start—a group of amateur gentlemen scientists. As such it enhanced the social respectability of science:

> Its membership lists, profusely decorated with the names of (mainly inactive) bishops, statesmen and aristocrats, enjoyed wide circulation,

[14] Ibid. 77. [15] Ibid. 34. [16] Ibid. 38.

disseminating esteem for the Society among non-members at home and abroad. The lists obviously gave the impression that the new science had been espoused by the establishment to an extent that was not necessarily the case.[17]

Even if the Royal Society overestimated its social sponsorship, it remained the case that many of the scientific men of the day came from wealthy aristocratic backgrounds. They had the educational opportunities and sufficient financial resources to spend on scientific experimentation. Science was carried on by men of private means in their leisure time; it was also produced by skilled artisans. Science was not yet a profession, nor did it bid for professional status. That was to come in the late nineteenth century. Meanwhile, men of law joined the ranks of the Royal Society. One lawyer, Sir John Clayton, was involved in a pioneering scheme for constructing lighthouses.[18] This is of interest to us, for it was the evidence of an engineer much concerned with the construction of lighthouses that the law first recognized as expert in *Folkes* v. *Chadd*. The membership of the Society suggests that lawyers and engineers were part of the same limited social circle. This fact may have accelerated the social currency of the new sciences. Mr Smeaton featured prominently on the cover of the first *Mechanics' Magazine*, the journal of the Mechanics' Institute movement, in 1823. He came from a middle-class background; he was the pioneering expert witness in *Folkes* v. *Chadd*; he was the first representative of the *nouveaux riches* and their science to gain public accreditation in the courts. At one level, Smeaton's appearance in court as an expert was not an especially radical departure; his testimony fell within known boundaries; he was commenting upon matters concerning the building of harbours, the course of tides, shifting of sands, and the like. Such expertise was familiar to the Admiralty Courts, where the judiciary regularly sat with trusted and known experts in such nautical matters.

Other new men of science were exploring quite novel territory. Theirs was new work carried out away from the public gaze in the mysterious confines of that new invention, the lab-

[17] Hunter, *Science and Society*, 48. [18] Ibid. 71.

oratory. Moreover, they claimed that their methods and techniques revealed objects, forces, and particles previously invisible to the naked eye. The new experts were asking the courts to believe things which judges and juries could not see, feel, hear, or touch. The reluctance of the judiciary to take these claims on board may have been due to the fact that these new experts were not gentlemen scientists but skilled artisans. In a society in which status and honour were ascribed rather than achieved, one could not simply rely upon the word of such men. With the expansion of science and technology in the nineteenth century, the number of new experts increased; the law could no longer be sure of the pedigree of the experts brought before them. Moreover, since these new experts were hired by the parties, and not court appointed, the law had little direct control over their selection. What control it did have, it had to exert upon them in the witness-box.

It was not only the rising middle class which had an interest in science. Some members of the landed classes had developed mining concerns on their estates, others were developing their agricultural stock, and still others were engaged in mercantile, industrial, and business concerns. They were interested in the application of scientific knowledge, in technological innovation rather than grand scientific theory. On one view, *Folkes* v. *Chadd* was a recognition by the law that it needed to take account of the new interests of its clients. It established the legitimacy of expert testimony in the law courts and recognized the currency of science in society more generally. Hunter also points out that, lower down the social scale, natural philosophy found an audience amongst artisans and merchants 'whose business had some scientific connection: manufacturers of scientific instruments, almanac makers and apothecaries'.[19] Restoration scientists were obsessed by the usefulness of their studies.[20] The Royal Society itself developed a project which would catalogue the history of technological processes used in the various trades. Hunter draws attention to the fact that many Fellows of the Royal Society became involved in trials in

[19] Ibid. 75. [20] Ibid. 87.

naval technology whilst others concentrated upon developing improved means of navigation and time-keeping across the globe. The Royal Observatory at Greenwich was founded in this period, with a view to perfecting navigation and astronomy. These were directly related to English colonial expansion by sea.

Experts demonstrated, in a very concrete manner, that they could manipulate nature and fashion man's environment; they could literally shape a new order of things. When invention was matched by incentive and capital, 'Practicks' worked.[21] But their combination could also pose a threat to the natural order of things. Rational planning in pursuit of the national good had reformist political implications.[22] If the world was nothing but a giant machine, the place of God as Chief Engineer stood in question. The new sciences threatened the established order and the established professions such as medicine, law, and divinity. They provoked charges of atheism, their heterodoxy challenged orthodoxy.[23] Science was heretical. The claim that nature is uniform was mirrored in the 'claim of both emerging capitalism and the liberal political theory that all men are equal'.[24] Such a claim was reflected in rule of law ideology, which promoted the view that all men were equal before the law, a view which Sachs describes as equalizing the unequal.[25] Changes in procedure were matched by developments in the substance of law as it attempted to meet the needs of emerging business and commercial interests for a system of law which was both rational and stable, but which limited state intervention in the market-place. Predictability was held to be a necessary virtue of rule of law, enabling individuals and groups 'to order their affairs with a confident look to the future'.[26] The loss of rule of law predictability threatened a descent into arbitrariness and anarchy, conditions under which the market could not survive and thrive.

[21] Ibid. 110. [22] Hunter, *Science and Society*, 110. [23] Ibid. 138.

[24] S. Harding, *The Science Question in Feminism* (Open University Press, Milton Keynes, 1986), 255.

[25] A. Sachs, 'The Myth of Judicial Neutrality', in P. Carlen (ed.), *The Sociology of Law* (Keele University, 1976), 110.

[26] Ibid. 111.

This project of modernity was, Foucault argues, a project of the bourgeoisie, 'the first class to claim universality'.[27] It was in this period that natural philosophers coined the term 'scientist'. They became the leading Grand Individuals of their day, leading British intellectuals with international reputations: 'It was they who were the self-confident militants of the period; [it] was they who were routing the theologians, confounding the mystics, imposing their theories on philosophers, their inventions on capitalists, and their discoveries on medical men.'[28] One of the inventions of science which now took hold in society was the post-Restoration distinction between value-free science and value-laden social and political beliefs. The methods as well as the findings of the New Science were to be impersonal.[29] This impersonal idiom, as Harding calls it, was a major site of contest between law and science during the latter part of the nineteenth century. Lawyers repeatedly sought to render the views of experts illegitimate by revealing the personal element in scientific inquiry. However, as Harding points out, unlike the socially legitimized persons of feudalism, objectivity and not subjectivity legitimized knowledge in bourgeois society precisely because it was a denial of privileged authority.[30]

The impersonal application of rule of law was mirrored by the impersonal application of rule by method. Those using such methods had equal access to the truth: 'Here science mirrors the hopes of liberal bourgeois man for an administrative form of ruling, a rule by procedure to replace personal rule by individuals.'[31] Paradoxically, those upon whom the law originally conferred legitimacy as experts were seen as highly individualistic social personages. However, since 'rule by method permits knowledge to be transferred from persons to things—from historical individuals to systems and machines',[32] the

[27] D. Ashley, 'Habermas and the Completion of Modernity', in B. S. Turner (ed.), *Theories of Modernity and Postmodernity* (Sage, London, 1990), 96.

[28] C. Chant and J. Fauvel, *Darwin to Einstein: Historical Studies on Science and Belief* (Longman and Open University Press, New York, 1980), 50.

[29] Harding, *The Science Question*, 228. [30] Ibid. [31] Ibid.

[32] Ibid. 229.

expertise of these superhuman sleuths gradually gave way to that of State-organized forensic science laboratories.

In a sense, then, the modern period is the period in which the aims, interests, and world-view of the bourgeoisie come to shape law as they shaped science. Developments in the jurisdiction of the courts, the formulation of rules of evidence, modes of trial, delineation of the duties of the prosecution and the rights of the defendant all represent a modernization of the legal system in the eighteenth and nineteenth centuries. This assisted in a programme of economic, political, and social modernization. At the same time, New Men and New Ideas of science emerged who were to 'tend to the health and welfare of the entire nation [and] . . . direct the life of the New Nature'.[33] Their articulation of science as an egalitarian activity in which both classes (the bourgeoisie and the proletariat) 'met on the fair field of science' served as a mechanism of social cohesion at a time of great social divisiveness and unrest.[34] The curriculum of the Victorian Mechanics' Institutes could be used to instil in the lower classes a sense that their position in society was simply a reflection of the natural order of things.[35] Coupled with the ideology of rule of law in which all men were equal before the law, science proved an exploitative resource for the bourgoisie generally and for law in particular.[36]

In fashioning the value-free version of what science was the new scientists were assisted by a commercial middle class with whose interests this coincided. For these merchants and industrialists to succeed in their bid for power, they required a mode of articulation in which their ascendance could be seen as the result of transcendental forces at work in the world. Since religious tradition had underpinned the divine right of kings to rule, and the right to own property had been ascribed

[33] F. Galton, *English Men of Science*, cited in Chant and Fauvel, *Darwin to Einstein*, 64.

[34] Burns (1837), cited in B. Barnes and S. Shapin, 'Science, Nature and Social Control, I: Interpreting Mechanics' Institutes', *Social Studies of Science*, 7 (1977), 31–74.

[35] Ibid.

[36] I. Cameron and D. Edge, *Scientific Images and their Social Uses* (Butterworths, London, 1979). See also D. Edge, 'Technological Metaphor and Social Control', *New Literary History*, 6 (1974–5), 135–47; Barnes and Shapin, 'Science, Nature and Social Control'.

to the aristocracy by birth, neither religious nor socially ascribed status could justify the rise of the middle classes. Paradoxically for these men busily engaged in the transformation of the material world by manufacture, neither would it serve to see any human hand directly on the tiller of destiny. By ascribing their social ascendancy to the working out of natural laws and a mastery over natural forces, the middle classes could claim that their ascendance was inevitable. The decline of the landed gentry was, by the same token, to be seen as an inevitability, the result of a natural selection and social evolution in which only the fittest survived. The invisible hand which shaped law and science thus also served to shape a new social order. It reproduced a social hierarchy which mirrored the newly discovered hierarchy of nature.

On one view, scientists were the intellectual carriers of this creed. Theirs was, they claimed, a vision unimpaired by social interests, their activities were unclouded by personal bias, their impersonal research methods allowed them simply to 'cast a mirror over nature'.[37] As the nineteenth century progressed, science was used in the law to advance the interests of middle-class industrialists and their commercial enterprise. It was the men of skills in technology, not men of pure science, who perfected the Factory System and the mechanized processes which made it the most profitable form for organization of labour. They not only provided new sources of profit and improved on old ones; they also provided various means of controlling the working classes. It was the new technology which de-skilled the Luddities and which demanded that workers become physically habituated to the discipline of the Factory System. It also began to be seen by legal utilitarians as providing the basis for a kind of scientific management of the dangerous classes who threatened to disrupt the social order. In this sense, the scientistic view of science did practical political as well as social

[37] M. Mulkay, *Science and the Sociology of Knowledge* (Allen & Unwin, London, 1979). See also B. Barnes, *Scientific Knowledge and Sociological Theory* (Routledge, London, 1974); B. Barnes (ed.), *Sociology of Scientific Knowledge* (Penguin, London, 1977); S. Woolgar, 'Irony in the Social Study of Science', in K. Knorr-Cetina and M. Mulkay, *Science Observed* (Sage, London, 1983); B. Latour and S. Woolgar (eds.), *Laboratory Life* (Sage, Beverley Hills, Calif., 1979).

work for the middle classes. Writing in the 1830s, Andrew Ure described, for example, how the division of labour brought about by new processes of spinning gave rise to a hierarchy within the factory, whereby some workers became subordinate to others.[38] These social relations of production ensured that discipline was enforced and the system managed without direct intervention of managers. They came to be seen as inevitable, that is, technologically determined. As parties to this process, engineers were shapers of social relations of production as much as they were the shapers of technology. In both capacities they contributed to the notion that technology 'may be seen as outside society',[39] that it is autonomous, neutral, and independent of social forces. Thus the inherently political and social nature of nineteenth-century technology and its social uses was effectively disguised. The questions asked and the knowledge developed to answer them were said to be determined by objects within the natural world itself rather than by social and cultural interests.

The middle classes and the landed gentry both latched on to various ideas expounded by the advocates of the new science. Some writers have argued that they did so for reasons of straightforward manipulation with a view to social control.[40] However, scientists began to make bigger and grander claims, namely that all other forms of knowledge must submit to science. This challenged the established order of things. John Tyndall, an important nineteenth-century physicist, had become a leading English scientist by the 1850s. Cowling writes that Tyndall took

> a high view of science and in pursuing a heroic, or romantic, conception of scientific leadership, addressed himself to two connected questions—the question of science as an investigative and regulative principle, and the question of nature as the object of science and the regulative context by which men had to be governed. Tyndall's science was an oligarchy of scientists who alone had authority in

[38] Andrew Ure, *The Cotton Manufacture of Great Britain*, vol. ii (1836).

[39] D. MacKenzie and J. Wacjman, *The Social Shaping of Technology: How the Refrigerator got its Hum* (Open University Press, Milton Keynes, 1985), 4.

[40] B. Barnes and S. Shapin, *Natural Order: Historical Studies in Scientific Culture* (Sage. London, 1986).

respect of Truth. But the most important aspect of scientific thought, and the whole of its merit as he saw it, was its method; it was the method, not the content of science, far less its consequences in terms of amenity of life, which gave science its significance as an instrument of culture.[41]

The tenor of Tyndall's address to the British Association, Belfast, in 1874 was enough to strike terror into the hearts of the traditional professions. Tyndall argued for the 'impregnable position of science' in contrast to all other 'schemes and systems which impinge upon the domain of science', theological or otherwise. They must submit to its control.[42] In the 1870s Tyndall was followed by Huxley, who expounded the majesty of fact, the primacy of scientific data based on evidence and reasoning, and the 'impossibility that men of science should recognise a theological reason and morality separate from scientific reason and morality'.[43]

This challenged traditional values and traditional sources of cultural leadership. If the law as an institution was to survive it had to be seen to take account of the developments brought about by science and technology. It sought to retrieve the value of the discipline by insisting that 'It is not always possible for man to arrive at a perfect knowledge of the truth in each particular case [and that] social necessities do not always allow him to suspend his judgment and refrain.'[44] Law, unlike science, was compelled to deliver an answer today. Unlike science, it could not allow its investigations to go on forever: 'Perhaps if we lived to an age of 1,000 years instead of sixty or seventy, it might throw light on any subject that came into dispute if all matters which could by possibility affect it were severally gone into, and inquiries carried on from month to month as to the truth of everything connected with it. I do not say how that would be, but such a course is found to be impossible at present.'[45]

[41] M. Cowling, *Religion and Public Doctrine in England* (Cambridge University Press, Cambridge, 1980), 144.

[42] C. A. Russell, *Science and Social Change 1700–1900* (Macmillan, London, 1983), 244.

[43] Cowling, *Religion and Public Doctrine*, 149.

[44] W. M. Best, *A Treatise on Evidence* (1849; Garsand, New York), 1976, 39.

[45] *A.-G.* v. *Hithcocks* (1847) 1 Ex. 91, 154 ER 38.

This was the 'great principle of the finality of judicial decision'[46] said to be universally recognized. On the basis of this principle, there must be a point at which judicial proceedings had to stop, otherwise, 'it would be productive of the greatest inconvenience and mischief if, after a cause civil or criminal has been solemnly decided by a court of competent and final jurisdiction, the parties could renew the controversy at pleasure, on the grounds of the real or pretended discovery of better arguments or evidence'.[47] Thus we have two mischiefs, one caused by delay and one caused by haste. If the promise of the new science was fulfilled, law could expect to be demoted to second place. What a relief, then, when it was discovered that men of science disagreed so frequently. This no doubt confirmed the belief of many nineteenth-century judges that the new experts were really a bunch of quacks and charlatans who needed careful watching. The function of the law was to keep the foundations of social life steady, building the pillars of the future on the foundations of the past. Scientists presented themselves as objective seekers after truth, an image which allowed the law to depict them as impractical men much given to the overturning of existing theories and, hence, the generators of uncertainty. Giving them too much power could even be construed as a threat to the social order. It would result in a loss of certainty. This, in turn, would result in the destruction of social—and economic -life which depended upon prediction and control. The court of law differed from the tribunal of posterity. It sought to deliver certainty. It must exclude evidence which would produce 'needless expense, vexation or delay'.[48] The proper discharge of its functions was that of 'expunging surplusage and restraining prolixity in pleading'.[49]

Another way of defusing the threat posed by the new experts was to incorporate them into the existing social structure. According to Russell, the scientific societies which some now see as a *locus* of nineteenth-century social control were, in fact, organizations primarily bound up with practical concerns,

[46] Best, *Treatise*, 41. [47] Ibid. [48] Thayer, *Treatise*, 43. [49] Ibid. 49.

only latterly taking on the aura of polite society and attracting social hangers-on. The majority of members of the scientific societies were in fact applied scientists involved in the commercial application of science and technology.[50] Their patrons, however, were often members of the aristocracy, who later hijacked the societies and made them into gentlemen's dining clubs where they could bask in the reflected glory of the scientific pursuit.[51] This private patronage of science was given as one reason why science did not require State funding.[52] Structurally, it also kept science within recognized and acceptable social boundaries. If experts wanted to be accepted they not only had to earn professional status, they had to acquire social status: they had to join the private club and accept its rules of membership.

This relationship between existing élites and upstart professionals was thus of benefit to each. The social status of experts increased by association, whilst the political élite secured an alliance which bestowed both financial and ideological benefits. Such a combination of interests probably explains the widespread currency of scientistic rhetoric in the nineteenth century. If we consider why the new scientists made these claims we must look at several contributory elements. Part of the answer lies in what Larson has called the professional project of the nineteenth-century professions. We can explain the claims of scientists in her terms as a seeking out of markets for their intellectual capital, seeking to transform intellectual resources into material and social rewards. Their knowledge was a commercial commodity with a market price. In the courts, the fee was set by the adversarial structure of legal proceedings. In the criminal courts, the only difference was that the leading experts were all bought up by the Crown, leaving the defendant to do as best he could in the market-place. Social status proved more elusive. It became necessary to gain the accreditation not only of one's profession but also of the Establishment; it was imperative to be accepted as a gentleman. The law, as part of the Establishment and one of what

[50] Russell, *Science and Social Change.* [51] Ibid. 196–7. [52] Ibid. 243.

Elliott has called the three status professions, had the power to confer this honour. The superior status of men of letters was in line with pre-existing intellectual traditions and cultural habits. Knowledge, and the means to knowledge, had long been the privilege of the ruling class. The new experts challenged this cultural supremacy. Part of the reason why experts were more readily accepted in the United States can be explained by the relative absence of such an entrenched cultural tradition. In England, there was a prevailing cultural snobbery which determined that men of letters could look down upon men of science. It was in this context that the new men of science made unrealistic claims. Critical to their bid for social and professional status was the acquisition of what Freidson has called 'gentry status'.[53] Gentry status brought with it connotations of disinterested dedication and learning, the tradition of the gentleman amateur.[54]

The coincidence of interests between ideas, economy, class, and power produced an idealized account of scientific work. Experts fashioned a strait-jacket for themselves and for future generations of expert witnesses. Their professional interests were best served by concealing the informal nature of scientific work.[55] Law, like science, could now pretend to be superhuman. Judges found the law, experts found facts; the facts were there to be uncovered and only by those specially equipped to interrogate the world. By insisting that they could distil fact from value experts fastened themselves on to the weakest of all legal structures, the fact/opinion distinction, and promulgated an idealized view of science. They opted for a view which best suited their short-term interests, the pursuit of social esteem, material rewards, and professional recognition.

[53] E. Freidson, 'The Theory of the Professions: State of the Art', in R. Dingwall and P. Lewis (eds.), *The Sociology of the Professions* (Macmillan, London, 1983).

[54] Ibid.

[55] See B. Wynne, *Rationality and Ritual: The Windscale Inquiry and Nuclear Decisions in Britain* (British Society for the History of Science, Chalfont St Giles, 1982).

5 The Rise of the Medical Detective

The Medical Police

As science expanded so the challenge to law continued. However, by the end of the nineteenth century several structural shifts had served to harness the professional expert's services to the agencies of law and order. Public health legislation, coupled with reforms of the Medical Act of 1858 and industry-specific demands for tighter laws on food adulteration, all generated a demand for public analysts, Poor Law medical officers, and public health inspectors. The concept of scientists as medical police was not new. In his 1798 lecture to the Patrons of Edinburgh University, Andrew Duncan discussed the function of the medical police. He saw their task in terms of applying the principles deduced from medical knowledge to the 'promotion, preservation and restoration of general health'.[1] Anxious to put his principles into practice, he encouraged the teaching of medical jurisprudence at Edinburgh: 'Many questions come before the Courts of Police where the opinion of medical practitioners is necessary either for the exculpation of innocence or the detection of guilt . . . There is no branch of medical education from which the practitioner may not derive useful information on some points necessary for enabling him to deliver before the Courts of Law, an opinion consistent with truth and justice.'[2]

Duncan's views on this matter prompted government ministers to consider the utility of medical jurisprudence and the expert witness:

On his answers the fate of the accused person must often depend, both Judge and Jury regulating their decision by his opinion. On the

[1] T. R. Forbes, *Surgeons at the Bailey* (Yale University Press, New Haven, Conn., 1985), 8.

[2] A. Duncan, Heads of Lectures on Medical Jurisprudence (Edinburgh, 1795), cited in Forbes, *Surgeons at the Bailey*. See also H. A. Husband, *The Student's Handbook of Forensic Medicine and Medical Police* (Livingstone, Edinburgh, 1874).

other hand, while he is delivering his sentiments, his own reputation is before the bar of justice. The acuteness of the gentlemen of the law is universally acknowledged . . . How cautious must, then, a medical practitioner be, when examined before such men, when it is their duty to expose his errors, and to magnify his uncertainties, till his evidence seem contradictory and absurd?[3]

Duncan's lectures at Edinburgh in this area and his concern with public health appear to have been part of a tradition which led to Edinburgh spawning a number of nineteenth-century medical reformers. Dugald Stewart's lectures in moral philosophy and political economy apparently influenced the thinking of leading politicians who studied under him at Edinburgh, including Palmerston, Russell, and Brougham.[4] One of his students, William Alison, later held the Chair in Forensic Medicine at Edinburgh for a brief time, before going on to become Professor of Medical Education and Physician to the New Town Dispensary in Edinburgh. Alison's *Observations on the Management of the Poor in Scotland and its Effects on the Health of the Great Towns*, published in 1840, was a major influence on public health reform in Scotland, mirroring the impact of Chadwick's influential *Report on the Sanitary Condition of the Labouring Population of Great Britain* on public health reform in England.[5] He was later to chair the central authority for public health matters in Scotland, the Poor Law Board of Supervision.[6] In Glasgow, Richard Millar, Professor of Medicine at Glasgow University, also produced work on the link between poverty and disease.

Elsewhere, medical men such as Southwood Smith proved influential advocates of public health reform. Smith had studied at Edinburgh, was a friend of Bentham and made a major contribution to Chadwick's work on reform of the Poor Law. He later joined Chadwick and Lord Ashley as the members of the newly established Central Board of Health, set up by the

[3] Memorandum from Henry Erskine to the Lord Advocate (1807), cited in Forbes, *Surgeons at the Bailey*, 7.

[4] M. W. Flinn, introd to E. Chadwick, *Report on the Sanitary Condition of the Labouring Population in Great Britain*, ed. M. W. Flinn (1842; Edinburgh University Press, Edinburgh, 1965).

[5] Ibid. [6] Ibid.

Public Health Act of 1848. The work of James Currie in Liverpool led to the compilation of detailed local surveys and laid the groundwork for the later inauguration of local Medical Officers of Health. Liverpool used the provisions of the Building Act 1846 to appoint the first of these. Interest in environments began to extend beyond their influence on the physical health of the public. Dr Thomas Clouston, the first official Clinical Lecturer in Mental Disease at Edinburgh in the 1870s,[7] stressed the new subject of mental hygiene. He aimed at educating the public on prudent and prophylactic life-styles[8] and argued that, like physical health, mental health might be encouraged by favourable environments. Poverty, destitution, malnutrition—complex economic and social factors could cause mental disease. Other causes included leading a dissolute life-style.[9] This analysis stressed the relationship between a healthy mind and a healthy body. If mental disease was attributable to alcoholism, pubs should be closed and the population should become teetotal; if syphilis caused mental deterioration, promiscuous sexual behaviour should be unlawful. Legislation such as the Contagious Diseases Act in the 1860s took these precepts to their logical conclusion. They constructed a causal link between environmental influences and deviance, the insanity and moral imbecility of the criminal classes and the areas they inhabited. It became necessary to map out the thieves' quarter, the rookery, and the areas of promiscuous assembly. Ecological theories of disease were extended to include moral and social diseases. Each branch of the public health endeavour aimed to manage the dangerous classes and their disease-ridden habitats. Just as medical reformers sought to eliminate typhus and cholera by eliminating squalor, bad housing, and poor sanitation, disciples of Thomas Clouston sought to improve the public's mental and moral hygiene; using forensic science to take criminals off the streets was simply an extension of the same public hygiene principle.

[7] M. Thompson, 'The Wages of Sin: The Problems of Alcoholism and General Paralysis in Nineteenth Century Edinburgh', in W. F. Bynum, R. Porter, and M. Shepherd (eds.), *The Anatomy of Madness: The Asylum and its Psychiatry* (Routledge, London, 1988), 316.

[8] Ibid. 317. [9] Ibid.

The involvement of medical men—including leading forensic scientists—in the various areas of public health was thus considerable. Flinn argues that the influence of the Edinburgh medical school extended beyond Scotland, since many medical reformers in England had trained at Edinburgh, where an emphasis was placed upon setting medical problems in their social context. This led them to an interest in the social origins of disease, and resulted in their producing a number of detailed local surveys concerning the social conditions in which diseases such as typhus and tuberculosis thrived. In a sense, then, Edinburgh appears to have been the home of the Medical Police, just as it was later to become the home of the Medical Detective.[10] One theory for the pervasive influence of Edinburgh-trained medical men is that they found a market for their medical services in the rising middle class and the newly organized corporations.[11] Flinn notes that the number of medical graduates from Edinburgh reached a peak in the 1820s, when 1,139 medical men graduated.[12] These doctors had received quite a new type of medical education, 'involving the integration of a wide range of medical and allied subjects . . . These well-trained doctors could by no means secure an adequate livelihood simply by treating the wealthy.'[13] Flinn argues that the production of a generation of like-minded medical men from Edinburgh combined with a rising interest in survey and statistical evidence to produce a movement for public health reform. This was driven primarily by doctors, professional administrators, and civil servants. As part of its bid to improve the status of the medical profession, the Provincial Medical and Surgical Association, founded in 1832, also engaged in medico-political and medical reform objectives.[14]

The end result was a new army of professional administrators, including factory and education inspectors, assistant commissioners, Poor Law medical officers, and local as well as central government civil servants.[15] The Medical Registration

[10] See J. Robertson, *A Treatise on the Medical Police* (Edinburgh, 1829).
[11] Flinn, introd. 18. [12] Ibid. [13] Ibid.
[14] N. Parry and J. Parry, *The Rise of the Medical Profession* (Croom Helm, London, 1976), 105.
[15] Flinn, introd. 18.

Act of 1858 also marked an important phase in the establishment of a unified medical profession[16] and the Royal Sanitary Commission of 1868–71 allowed doctors to spearhead the debate on sanitary reform. The Poor Law infirmaries provided career opportunities for doctors in the public sector. This also gave them an organized voice in national debates about the running of Poor Law workhouses, infirmaries, and asylums.[17] In the 1860s, John Simon, Medical Officer of the Privy Council, conceived the State's role as superintendent-general for health, monitoring sanitary reform and providing the foundation for the later development of a class of State physicians.[18] These medical superintendents and public service supervisors provided a new type of disciplinary gaze which extended rather than diminished the power of law.

Medical men thus had a role in the supervision and regulation of the population. The mid-1800s also saw the rise of a number of institutions concerned with the production of population statistics, mortality returns, and the civil registration of births, deaths, and marriages. In part, Flinn argues, this was stimulated by legislation requiring that the ages of factory children be authenticated, and by the demands of Nonconformists for a legally valid means of registering births, deaths, and marriages which would be independent of the Church of England.[19] It was also stimulated by the work of the actuaries such as Joshua Milne, employed by the Sun Alliance Assurance Society to compile figures on life expectancy.[20] John Finlaison, an actuary working in the National Debt Office, produced the first comprehensive statistics on life expectancy, mortality, and public health, and called for a system of registration which would enable administrators to monitor public health and population trends. Local institutions such as the Manchester Statistical Society and the London Statistical Society also contributed to the rise of these vital statistics as a form of information and a basis for public policy.

This work was furthered by William Farr, a doctor who

[16] Parry and Parry, *The Rise of the Medical Profession*, 131.
[17] Ibid. 141.
[18] Ibid. 154.
[19] Flinn, introd., in Chadwick, *Report*, 26 ff.
[20] See ibid. 12.

became Compiler of Abstracts in the newly established Registrar-General's Office in 1837. Farr had published his 'Lectures on Hygiene or the Preservation of Public Health' in the *Lancet*,[21] and also his article 'Vital Statistics' in J. R. McCulloch's *A Statistical Account of the British Empire* in 1837.[22] Flinn argues that, like civil servants in other government departments, Farr used the statistics included in the Annual Reports of the Registrar-General's Office to push a particular cause, in his case medical reform and public health.

The involvement of the new scientists and medical men in public policy debates about medical and social reform thus took place on three levels. There were those actively involved in improvement of certain towns and cities, for whom the relationship between poverty and various forms of disease became self-evident. Others sought to back up this argument with statistical information as a basis for national policy on public health. And there were those engaged in university teaching whose lectures on disease widened the context of discussion to include social and economic factors. The leading forensic scientists and medical jurisprudence professors of the day incorporated public health concerns into their own work. Alfred Swaine Taylor wrote about the public utility of forensic science. Forbes tells us that John Gordon Smith considered a concern with public health and with the administration of public institutions for the cure of diseases as one specific branch of forensic science which he termed the Medical Police. Traill's *Outlines of a Course of Lectures on Medical Jurisprudence* also includes a section on the Medical Police.[23] Forbes notes that Sir Henry Letheby became Medical Officer of Health for the City of London; he was also a physician, analytical chemist, professor of chemistry, food analyst, and chief inspector of illuminating gas for the City of London. In his capacity as protector of public health, he gave evidence in public health and safety cases. Both Taylor and Letheby gave expert evidence in an 1855 case concerning the hazards of storing wood naphtha in

[21] *Lancet*, 1 (1835), 36.

[22] Flinn, introd., in Chadwick, *Report*, 26–9.

[23] T. S. Traill, *Outlines of a Course of Lectures on Medical Jurisprudence* (Black, Edinburgh, 1836). See also Forbes, *Surgeons at The Bailey*, 207.

a residential area. Letheby also testified in cases concerning the sale of spoiled and diseased meat. These cases came at a time when the *Lancet* was campaigning for the control of the sale of such foods by law.[24] The job of the forensic scientist was extended to include the tasks of public analyst, monitoring the adulteration of foodstuffs, and aiding prosecution of offenders.

The concern with public health reform formed part of the more generalized Victorian push for social purity and turn-of-the-century concerns with the need of the nation for a healthy and productive citizenry. The growth of hospitals, orphanages, asylums, and prisons has been linked theoretically by writers such as Foucault, Scull, Donzelot, and Piven and Cloward with the extension of State control and the regulation of the population by, amongst others, medical men, psychiatrists, and social workers.[25] Flinn's analysis of public health reform also stresses the administrative regulation of persons by professional administrators. It is in this context of developing means of social control at a time of immense social and economic flux that the growth of a professional police force occurred, and with it the development of State-funded forensic science. Forensic science concentrated upon physical remains and traces of bodily contact between a place, a victim, and a suspect. It was concerned with anatomy as a technique of surveillance. At one level this mirrored the political anatomy of generalized Panopticism, and was part of the centralized registration of the pathological in disciplinary society.[26] An Act of 1884 further contributed to this alignment. This Act redefined the office of Treasury Solicitor and established the office of Director of Public Prosecutions. In 1908 it was followed by the Prosecution of Offences Act and in 1907 the Court of Criminal Appeal was established. It was under the auspices of these provisions that experts were called in by the Crown to

[24] Forbes, *Surgeons at the Bailey*, 209.

[25] See e.g. M. Foucault, *Discipline and Punish* (Pantheon, London, 1975); M. Foucault, *Madness and Civilization* (Pantheon, New York, 1965); A. Scull, *Museums of Madness* (Allen & Lane, London, 1979); F. Piven and R. A. Cloward, *Regulating the Poor* (Tavistock Press, London, 1972); J. Donzelot, *The Policing of Families* (Pantheon, New York, 1980).

[26] Foucault, *Discipline and Punish.*

assist with the growing volume of criminal prosecutions which had been generated by the establishment of professional police forces. The last two decades of the nineteenth century also saw the development of improved methods of identification, such as those developed by Bertillon in 1886; this was superseded in 1892 by dactyloscopy, which was a system based on the uniqueness of fingerprinting. Forbes notes that systematic identification of criminals by their photographs was first suggested in 1894 by the Secretary for the Home Department.[27] Taken together, the plethora of new regulations and statutes provided a wider framework for the application of forensic scientists.[28] Of the leading forensic scientists of the day, Luff, Pepper, Wright, and Wilcox became the experts retained by the Home Office. Pepper was a founder member of the Medico-Legal Society and, with these others, comprised the elite of 'by far the most prominent and influential figures in forensic science'.[29]

The Medical Detectives: Taylor, Spilsbury, Luff, Wright, and Willcox

If public health officials and public analysts were the medical police, Crown forensic experts were to become the medical detectives. They achieved a high profile in a series of much-publicized cases, and in turn were rewarded with unimpeachable authority in the law courts and high social status. The man of law was a brilliant and mercurial genius; the man of science was aloof and respected. Cultural images of the man of science increasingly depicted him as superhuman. His detective skills drew on a vast and encyclopaedic knowledge in the arts and the sciences. He was acquainted with the nobility of England and mixed with them as an equal. His forensic expertise was seen as a combination of art and skill, and therefore something more than a mere vocation. He was, indeed, an amateur gentleman detective rather in the mould of the ama-

[27] Forbes, *Surgeons at the Bailey*, 218.
[28] D. G. Browne and E. V. Tullett, *Bernard Spilsbury* (Harrap, London, 1951), 36.
[29] Ibid. 26.

teur gentleman scientist. This formula has helped shape modern cultural expectations of the forensic scientist. The mould was set by the type of experts who found approval in the Victorian and Edwardian courts, and was epitomized by experts such as Sir Bernard Spilsbury and Alfred Swaine Taylor. They were not upstart professionals. They were society's watchdogs of the night. Indeed, the *Lancet* described Bernard Spilsbury as 'a pillar of security',[30] whilst others lauded him as 'the obvious prototype of the doctor-detective of modern fiction'.[31] The work of such men as Taylor, and later of the so-called Willcox Circus,[32] captured the public imagination and lent the law immense public credibility. These were the experts who supplied the means to trap even the most sophisticated villain. They could not only prove him guilty beyond reasonable doubt; they could prove him guilty beyond any doubt at all.

The professional status of these new scientists was wrought by a variety of means and after several struggles with the law. Russell cites a case from 1820[33] which centred on liability for losses in a fire at a firm's Whitechapel premises:

> Many chemical witnesses were called, including W. T Brande, J. T. Cooper, F. C. Accum, T. Thomson and Michael Faraday. Since a witness had to have professional status in order to claim expenses, this became a matter of further contention. In November 1821, the court ruled that a man of science, without recognised learning and actually basing arguments on the novel process of making experiments was no more professional than a mechanic. In thus denying science equivalent status to law, medicine or the Church, the judges were reflecting the established social values of the ruling class . . . The decision settled for many years the status of a chemist, despite the fact that chemists frequently made large sums of money as expert witnesses in litigation.[34]

[30] *Lancet*, (27 Dec. 1947), 965.

[31] *DNB* (1941–50), 815, 'Spilsbury'.

[32] The Willcox Circus is also sometimes known as Wright's Circus after Sir Almroth Wright, bacteriologist and mentor to the group.

[33] *Severn King and Co.* (1820) CCp., cited in C. A. Russell, *Science and Social Change 1700–1900* (Macmillan, London, 1983).

[34] Russell, *Science and Social Change*, 221.

Law played an important part by granting or withholding professional recognition to the experts appearing in its courts. In this sense, it was the consumers of expert services who exercised control over their professionalization. The career of Sir Alfred Swaine Taylor provides a good example of how State experts were elevated above those appearing for the defence. Taylor was an eminent forensic scientist; until his death in 1880, he was the leading Treasury expert. He was Professor of Chemistry at Guy's Hospital and the author of the leading text on medical jurisprudence. He moved in high circles. Indeed, in the Smethurst trial in 1859, counsel objected to the judge in the case, Sir Jonathan Pollocks, Lord Chief Baron of the Exchequer, because he was a close personal friend of Taylor. The judge's reply was that 'almost any judge who belonged to a literary or scientific society might be found to be acquainted with the well-known Dr Taylor'.[35]

Taylor enjoyed the courts' protection for most of his career. His speciality was the analysis of poisons and he appeared for the Crown in almost all of the celebrated poisoning trials of the late nineteenth century. Poison in small quantities was virtually impossible to detect; hence its detection required extraordinary skills. Experts in poisoning cases acquired a high public profile. Orfila had 'attained the fame or notoriety of a modern movie star through his appearance in the great poison murder trials of the period'.[36] Orfila had written his *Traité de toxicologie* in 1813; Sir Robert Christison wrote *A Treatise on Poisons* in 1829, and Taylor wrote his book *On Poisons* in 1848. Taylor was England's equivalent of Orfila, closely followed by Sir Robert Christison. Christison held the Regius Chair of Medical Jurisprudence at Edinburgh, a position he had obtained partly with the help of ministerial and Crown patronage in Scotland. His most famous case had been that of Burke and Hare, the so-called body-snatchers, in 1829. He was called to give evidence in the trial of William Palmer, a trial which

[35] D. H. Knott (ed.), *The Trial of William Palmer*, rev. E. R. Watson, Notable British Trials Series (1856; Hodges & Co, London, 1952).

[36] E. H. Ackernecht, 'The Early History of Legal Medicine', in C. R. Burns (ed.), *Legacies in Law and Medicine* (Science History Publications, New York, 1977), 263.

the presiding Lord Chief Justice called the most memorable judicial proceedings for the last fifty years, engaging not only the attention not only of this country but of all Europe.[37]

The defence experts ranged against Taylor and Christison at Palmer's trial in 1856 faced an uphill struggle. They were challenging the renowned and revered experts of the day. Taylor admitted in his opening evidence that he had never before had experience of the action of strychnia on the human subject, but that he had written a book on the subject. He had tried various experiments on rabbits, in which post-mortem examination had failed to reveal that strychnia had been administered. The defence argued that although the Crown experts were diligent men, their failure to find any trace of poison meant that their belief that it had been administered could only be a hypothesis. To support this contention, the defence called its own experts including Dr Nunneley, Fellow of the Royal College of Surgeons and Professor of Surgery at the Leeds School of Medicine; a Dr Williams, Professor of Materia Medica at the Royal College of Surgeons in Ireland; Dr Henry Letheby, Professor of Chemistry and Toxicology at the London Hospital Medical School; Dr Robinson of the Royal College of Physicians; Dr Nicholas Parker, Professor of Medicine at the London Hospital; Mr Rogers, Professor of Chemistry at St George's School; and lastly, the most eminent chemical analyst in the country, Mr William Herapath of Bristol. Herapath totally rejected the prosecution's theory as utterly unworthy of credence.

Dr Nunneley opened the scientific evidence for the defence. He was a member of many British and foreign scientific societies. In his opening evidence he testified that he had read the depositions of the other witnesses and disagreed with them. Lord Campbell immediately objected to this, arguing that any disagreements must be founded upon the viva voce depositions of these witnesses during the trial. When Dr Nunneley began again, basing his views first on his own findings, then on the viva voce statements of previous witnesses, he was immediately

[37] Knott, *Trial of William Palmer.*

interrupted, Lord Campbell announcing that medical men were not to be substituted for the jury. Dr Nunneley was then asked to set out each of the facts he had assumed in arriving at his conclusions. After a number of other interruptions from Lord Campbell, Dr Nunneley was succeeded in the witness box by Dr Herapath, Dr Rogers, and Dr Letheby. In his opening evidence, Dr Letheby revealed that he had been employed by the Crown in almost every poisoning trial during the last fourteen years. He testified that it was possible to detect strychnia and that he had done so on several occasions.

Despite the soundness of their arguments and their high level of expertise, the defence experts were dismissed. The Attorney-General described Letheby as being strongly biased in favour of the defence. Of Dr Nunneley he said, 'He certainly seemed to me to give his evidence in a manner not quite becoming a witness in a Court of justice.'[38] Dr Herapath, he argued, had said that if strychnia had been present Dr Taylor ought to have been able to find it: 'He intimated . . . that there might have been strychnia, and that Dr Taylor did not use the proper means to detect it.'[39] Dr Herapath came in for particular criticism from the presiding judge: 'Mr Herapath has again and again asserted that this was a case of poisoning by strychnia, but that Dr Taylor had not known how to find out. Yet he has been prepared to come forward as a thorough-going partisan, advising the defence, suggesting question upon question on behalf of a man who he has again and again asserted he believed to be a poisoner. I abhor the traffic in testimony in which I regret to say men of science sometimes permit themselves to condescend.'[40] Some, he said, were too concerned to secure the accused's acquittal, adding that a witness should not be turned into an advocate any more than an advocate should be turned into a witness: 'It is for you (the jury) to say whether some of those who were called on the part of the prisoner did not fall to the category which I described as witnesses becoming advocates.'[41] The court accused the defence experts of being motivated by money and/or profes-

[38] Knott, *Trial of William Palmer.* [39] Ibid. [40] Ibid. [41] Ibid.

sional jealousy whilst singling out the Crown experts for praise. The Attorney-General said he 'could not help saying that it seems a scandal upon a learned, a distinguished, and a liberal profession, that men should come forward and put forward such speculations as these, perverting the facts, and drawing from them unwarranted conclusions with the view of deceiving the jury'.[42] Given such a rebuke, it is perhaps little wonder that Sir Keith Simpson was later able to write that Taylor 'stood virtually unchallenged in his field for nearly forty years',[43] and Forbes was able to describe him as towering among the forensic scientists of his day.[44]

In July 1859, Taylor was once again the leading Crown expert in the trial of Thomas Smethurst. In this instance, he testified that he had examined fluids taken from the body of the victim, which he said contained arsenic. However, during the course of the case, he changed his mind, admitting that he had been mistaken. Later testing by a different method showed arsenic to be absent. Taylor discovered that the liquid was chlorate of potash, whose action upon the copper gauze resulted in arsenical impurities in the copper being released into the liquid. Taylor admitted in court that the equipment he had used to test the fluids might itself have adulterated it; he could not be sure whether arsenic had been present. Admitting this mistake did not, however, dissuade Taylor from his belief that arsenic had been administered to Isabella Bankes. Once again, there was public concern about a conviction based upon Taylor's evidence. This took the form of letters to the medical as well as to the popular Press, and to the Home Secretary. Eventually the matter was decided by Sir Benjamin Brodie, Queen Victoria's surgeon. He found six reasons for thinking Smethurst guilty and eight that he was perhaps not guilty. It was not, therefore, possible to say that Smethurst was guilty beyond all reasonable doubt. Objection was raised on the grounds that 'The immediate result of this move was . . . that the responsibility of the decision was thus shifted from those on whom it properly rested on to a man

[42] Ibid. [43] K. Simpson, *Forty Years of Murder* (Panther, London, 1980).
[44] Forbes, *Surgeons at the Bailey*, 139.

who, however skilled and learned as a surgeon, was neither a juryman nor a judge.'[45] The outcome was a free pardon on the Home Secretary's recommendation.

Since it was Taylor's reputation rather than his experimental wizardry which made him such a crucial plank in Crown prosecutions, the Treasury began to employ some of Taylor's fiercest critics, thereby removing from the defence most of the available experts. There was a gentleman's agreement that experts employed by the Crown did not offer their services to the defence. Henry Letheby and Dr Bronte were two experts who did not abide by this agreement. Their frequent appearances for the defence were seen as a defection. The cooptation by the Crown of most of the country's leading experts gave it a virtual monopoly on forensic expertise. In a *laissez-faire* world, it was only the defence which had to secure its experts in the free market.

Spilsbury: The Making of a Legend

By the end of the nineteenth century the Crown had secured to itself the services a formidable group of forensic experts, most of whom belonged to the so-called Willcox Circus. It was from within this group that Taylor's successor was found. As with Taylor, Sir Bernard Spilsbury's currency as a Crown witness grew to the point where he became virtually unimpeachable. When he appeared for the defence in a Scottish case the defence were said to be fortunate indeed to recruit an expert of Spilsbury's calibre, 'of European fame, of the highest eminence in his profession', whilst Craigie Aitchison, appearing for the defence, called him 'St Bernard'.[46] Spilsbury came to command great respect in the courts. He was dubbed the Sherlock Holmes of forensic science, securing 'the convictions of more murderers than any other man working in the cause of justice'.[47] His public image was of a man 'impartial and sci-

[45] Ibid.

[46] Browne and Tullett, *Bernard Spilsbury*, 86.

[47] *Times Literary Supplement*, 21 Nov. 1936 (review of L. Randall, *The Famous Trials of Sir Bernard Spilsbury*).

entifically reasonable, difficult to tilt, aloof and a little cold'.[48] He came to epitomize the positivist ideal of what a man of science should be. The high public profile of Spilsbury's trials frequently placed him in the headlines. He became well known to the public, gathering to him a reputation for snaring the most clever and wicked of villains, a reputation which to some extent still persists.

Franco Moretti argues that the courtroom dramas in which the likes of Spilsbury featured as folk heroes were quite in keeping with wider social developments. In the nineteenth century, the focus of attention shifted from the public execution to the trial and, 'While the former underlines the individual's weakness by destroying his body, trials exalt individuality.'[49] They exalted the individuality not only of the villain, but of also of the lawyer, the judge, and the expert witness. The courtroom drama was enacted by notable players in their assigned roles, and being able to put a name to a face gave added comfort in a society which was becoming increasingly anonymous. Moretti argues that detective fiction of the time also emphasized an individualistic ethic. In his view, the dangerously nonconformist element of individualism was, however, no longer associated with the middle classes:

> Reading Conan Doyle, however, one discovers that criminals are never members of the bourgeoisie. The bourgeoisie is no longer the champion of risk, novelty and imbalance, but of prudence, conservation and stasis. The economic ideology of detective fiction rests entirely upon the idea that supply and demand tend quite naturally towards a perfect balance. . . . The criminal generally belongs to one of two major sociological types: the noble and the upstart.[50]

In the stories of Sherlock Holmes, Conan Doyle writes of a society beset by secret societies, the formidable power of the Masons, the sinister activities of trade unions, and the dastardly clever pursuits of scientific criminals. The normal forces of law and order are no match for their intelligence. Only a

[48] *Lancet* (27 Dec. 1947), 965.

[49] F. Moretti, *Signs Taken for Wonders: Essay in the Sociology of Literary Forms* (Verso, London, 1983), 138.

[50] Ibid. 139.

remarkable individual can stop them. The motives of such a detective are altruistic. His ends are the pursuit of scientific detection and the good of society. This accords with the ethos of professionalism, which stresses personal integrity, altruism, service in the public interest, and so on. In this context, a gentleman's word is his bond—it was thus essential that those who aided the forces of law and order be deemed gentlemanly, professional scientists.

The detective work of fictional heroes such as Sherlock Holmes and Lord Peter Wimsey, and accounts of factual heroes such as Sir Bernard Spilsbury, were directly bound up with the purposes of establishing control over a shifting social order. The latter half of the nineteenth century had seen a revival of social and political unrest, and a unparalleled growth in the population of the towns and cities of England. Whereas the police did practical work at the level of investigating and detecting crime, forensic experts also did symbolic work, creating the impression that, once the power of the scientific vision had been harnessed to law, not even the cleverest of villains could escape conviction. Whilst Spilsbury was at work, clever criminals would find no hiding place:

> Until the beginning of this century, a clever murderer had a good chance of getting away with his crime in Britain. Scotland Yard detectives were expert at catching a culprit but proving his guilt in the meticulous British courts was far more difficult. Sir Bernard Spilsbury, by introducing the exact science of the laboratory into the courtroom, changed that. 'I hate murder', he once said, 'and I have devoted my life to making that most unforgiveable of all crimes unprofitable' . . . The quiet, modest scientist, for 35 years the principal pathologist for the Home Office, was the father of modern legal medicine. His evidence was instrumental in convicting more than 100 dangerous killers.[51]

In its obituary upon Spilsbury's death in 1947, *The Times* commented as follows:

> between the wars there was hardly an instance in which Scotland Yard, requiring the painstaking collection of medical evidence to

[51] *Reader's Digest*, 55 (Sept. 1949).

convict a murderer, did not benefit from his [Spilsbury's] unique presentation of the facts. As a witness he never said a word too much and juries were invariably impressed by his clear-cut way of giving evidence, and his unmistakeable authority of statement . . . It would be difficult, if not impossible, to cite one case of poisoning or of dismemberment of a victim in which his skill failed to establish the guilt of the accused person. As his reputation grew, he figured more and more in the fierce glare which always surrounds trials for murder and was involved in nearly all the famous cases tried at the Old Bailey and elsewhere.[52]

The motif of the intelligent gentleman detective appeared repeatedly not only in fiction but also in the Press and in the personal biographies of experts. The detective of this idiom has a kind of Panoptical vision in which members of mass society are at once rendered observable, traceable, and controllable. He is, as Moretti says, 'the figure of the state in the guise of "night watchman", who limits himself to assuring respect for laws'.[53] Fictional detectives like Sherlock Holmes 'know all the possible causes of every single event. Thus the relevant causes are always a finite set. They are also fixed . . . Holmes cannot go wrong because he possesses the stable code . . . [he] takes in the only possible meaning of the various clues in a glance.'[54] He is

the great doctor of the late Victorians, who convinces them that society is still a great organism: a unitary and knowable body. His science is none other than the ideology of the organism: it celebrates its triumph by instantaneously connecting work and exterior appearances (body, clothing): in reinstating an idea of status society that is externalized, traditionalist, and easily controllable . . . He degrades science: just as it had been humiliated by both the English productive structure and the education system as the turn of the century. But at the same time, he exalts it. The need for a myth of science was felt precisely by the world that produced less of it. England did not attain the second industrial revolution: but it invented science fiction.[55]

[52] *The Times*, 18 Dec. 1947.
[53] Moretti, *Signs Taken for Wonders*, 155.
[54] Ibid. 145.
[55] Ibid.

Culturally constructed images of scientific detectives were powerfully reassuring. These quiet, unassuming scientific men had an ability to deconstruct that most threatening of all criminal types, the psychopath, whose unfeeling, impersonal, clever, and calculating villainy assaulted the very basis of social life. In the vast crowds of the new metropolis the criminal could be anyone; he could slip into the background and pass unnoticed in the crowd. Scientists had the means to undo this cloak of unsettling anonymity. What we observe here is one cultural stereotype undoing another. Medical detectives such as those in the Willcox Circus were described as quiet men whose diligent work kept the peace and enabled the citizens to feel safe in their beds. Whether they provided real security did not really matter. Of greater importance was their symbolic role in a society beset by rapid social, economic, and political changes. The felt need of the State was for extraordinary individuals to catch extraordinary villains. Crime and social disorder were everywhere; organized police forces partly contained the dangerous classes but they were no match for the dangerous individual. In the metropolis, there was a fear that social development might make control impossible. Moretti argues that detective fiction's answer to this was to make it impossible for the villain to hide within the crowd. However, it took a man with extraordinary powers to trace the minute clues he left behind.[56]

Crown experts themselves, as well as the general public, increasingly identified detective fact with detective fiction. From this blurring of boundaries emerged a strong cultural image of forensic science and forensic scientists. At least one fictional detective was modelled on Spilsbury whilst Sherlock Holmes, the most famous of all the scientific detectives, is generally held to have been based upon Dr Joseph Bell, an Edinburgh University medical lecturer. Bell himself had been a student of Sir Henry Littlejohn, Professor of Medical Jurisprudence at Edinburgh, later appointed first Medical Officer of Health for the city of Edinburgh, and father of

[56] Moretti, *Signs Taken for Wonders*, 155.

Professor Harvey Littlejohn, who succeeded him as Professor of Medical Jurisprudence. The interplay between science and fiction was strengthened by Conan Doyle's own involvement in public protests concerning convictions in several notable trials. Spilsbury's integrity was assumed, though Conan Doyle and others objected to the unquestioning faith placed in his testimony, arguing that 'For some reason or other, Sir Bernard Spilsbury has now arrived at a position where his utterances in the witness box receive unquestioning acceptance from judge, counsel, and jury. He can do no wrong. But a reputation for infallibility such as that which appears of late to have been thrust on Sir Bernard (I am sure he never claimed it for himself) is quite out of place in medical and surgical matters.'[57] How then did Spilsbury's evidence acquire such vast importance?

> The answer is simple. Juries are formed from members of the public, and the British public believed Spilsbury infallible . . . Spilsbury had indeed done what few can hope to do; he had become a legend in his own lifetime. To the man in the street he stood for pathology as Hobbs stood for cricket or Dempsey stood for boxing or Capablanca for chess. His pronouncements were invested with the force of dogma, and it was blasphemy to hint that he might conceivably be wrong.[58]

This image of Spilsbury seems to have been encouraged by his own tenacity, seen by his defenders as a virtue and by his detractors as a vice. Once he had made up his mind nothing could shake him. Some spoke ironically of Spilsbury's papal infallibility. He was described as 'a professional cross-examinee . . . as used as any counsel to the atmosphere of the courts, as trained in the rules, as familiar with the tricks'.[59] It was objected that juries took Spilsbury's word whatever the evidence of opposing witnesses simply because he had great authority. J. D. Cassells, QC, expressed a familiar fear that it

[57] Anonymous letter to the Press, cited in Burns (ed.), *Legacies in Law and Medicine*, 175.

[58] Ibid. See also E. Lustgarten, *Verdict in Dispute* (1949), cited in Browne and Tullett, *Bernard Spilsbury*, 176, 184–5, 263.

[59] Browne and Tullett, *Bernard Spilsbury*.

would be 'a sorry day for the administration of criminal justice in this land if we are to be thrust into such a position that, because Sir Bernard Spilsbury expressed an opinion, it is of such weight that it is impossible to question it'.[60]

Experts such as Taylor and Spilsbury set up expectations which were to prove a somewhat ambivalent blessing for those who came after. The image which they created was an idealized image of forensic science. By refusing to countenance the views of their contemporaries, they denied the plural nature of scientific inquiry. Spilsbury provided the first major articulation of what a forensic scientist should and could do. He helped the police. He was part and parcel of an apparatus which linked forensic science to the State and social control. Throughout Europe at the turn of the century, criminalistics grew as a discipline and as an aid to prosecution. The range of bodily features by which a suspect might be identified multiplied. It included bloodstains, hair, seminal stains, saliva—a panopoly of biological characteristics were appended to the State's machinery of control. A system of surveillance, of social, moral, and physical eugenics, began to be built up. In 1893 the first textbook on criminalistics was published in Germany; in 1907 it was translated into English. Locard began his Institute of Criminalistics in France in 1910. In England the individual men of science gave way to a full-time corpus of prosecution forensic scientists in 1935, with the establishment of the Metropolitan Police Laboratory, whilst in the USA the FBI Laboratory began in 1932.[61]

Co-optation into the élite

In the absence of an independent profession of pathology, and given the understanding that Crown experts would not testify for the defence in England, defence lawyers were forced to look elsewhere for someone to oppose the Crown's expert evidence. There were few such experts to choose from. Dr

[60] Cited in Simpson, *Forty Years of Murder*, 30.
[61] M. Green, 'Is Sir Bernard Spilsbury Dead?', in A. Brownlie (ed.), *Crime Investigation: Art or Science?* (Scottish Academic Press, Edinburgh, 1984), 23.

Bronte, an Irish pathologist, was one of these. In the account given by Browne and Tullett in their leading biography of Spilsbury, Bronte is described as 'the typical Irishman . . . clever, quick-witted, voluble, combative, sociable, possessed of the gift of making friends and partisans . . . But he was also pushing, self-opinionated, with little sense of dignity, and rather boastful . . . A graver complaint against Bronte was that he was slapdash in his work.'[62] Bronte was described as a 'so-called pathologist', and as rather careless. Denigrating defence experts in such terms was one means of preserving the credibility of Crown experts. The best of them were co-opted by the State. They were frequently invited to dine with the men of arts and men of law in their gentlemen's clubs; they formed learned societies such as the Medico-Legal Society and the Crimes Club, where lawyers and experts mixed with others interested in forensic matters. The founding of Our Society (the Crimes Club) itself grew 'out of an evening party at the house of H. B. Irving, the actor' and it was 'to include judges, counsel, coroners, doctors, writers, actors and others interested in crime'.[63] This type of social gathering generally centred around a few Grand Individuals, who discussed crime and the criminal rather as if they were some curious species from an exotic and unexplored continent: 'On certain evenings once every month a group of men and women, many of them famous names in current criminology, gather in a London building to take part in proceedings any student of real-life or fictional crime would find fascinating beyond compare. They are all members of one of the most interesting bodies in Britain—the Medico-Legal Society, which is the nearest thing to a serious Crimes Club the world of criminology possesses.'[64]

Through such social networks, certain experts came to be recognized by the élite. These few experts accomplished what Rueschemeyer has decribed as an anchoring of their claims in cultural, political, and socio-economic institutions.[65] They were

[62] Browne and Tullett, *Bernard Spilsbury*, 167. [63] Ibid. 31.

[64] S. Firmin, *Scotland Yard: The Inside Story* (Hutchinson, London, 1948), 131.

[65] See D. Rueschemeyer, 'Professional Autonomy and the Social Control of Expertise', in R. Dingwall and P. Lewis, *The Sociology of the Professions*, (Macmillan, London, 1983), 49.

seen as leading experts in their field, even though their professional colleagues might hotly dispute this claim. We have seen, for example, that Alfred Swaine Taylor's skills as an analyst were not highly rated by his fellow scientists, and though Sir Bernard Spilsbury obtained the law's stamp of approval, 'to many in the profession who spoke of pure medicine as an art, the study of morbid anatomy and pathology was still a "beastly science".'[66] The conferring of gentlemanly status upon Crown experts was one element of a process whereby they gained cultural authority in Victorian and Edwardian society. It must be understood in the context of a society which began to confer status in terms of achievement as well as birth. In his review of Larson's work, Halliday has argued that, in monopoly capitalist societies, professions bear 'the outmoded values of liberal capitalist formations: individualism, expertise, and status-seeking'.[67] Thus, 'the professions spawned by competitive capitalism come to sustain its monopoly successor, the most visible carriers of an illusion on which mature capitalism depends'.[68] Professions carry 'residues of a pre-capitalist past—ideals of intrinsic value in work, of universal service, of noblesse oblige'.[69] Most important of all, Halliday argues, is 'the irony that the professions, having conquered the peaks of occupational status, are the victims of their own ideology. They have no power of their own. They are merely the agents of power. Their vaunted distinctions of autonomy and control over their occupational activity in fact only conceal collective powerlessness, subordination, and complicity.'[70]

Experts such as Taylor and Spilsbury were carriers of this creed of professionalism but they obtained material and social rewards at the cost of professional autonomy. They shaped the predominant account of the forensic expert's role. He was primarily an employee of the State. His job was to help the police decide whether a crime had been committed, how, when, and by whom. He was to try and establish links between suspects and the scene of crime by adopting Locard's

[66] Browne and Tullett, *Bernard Spilsbury*, 23.
[67] T. Halliday, 'Professions, Class and Capitalism', *Arch. Euro. Socio.* 24 (1983), 328.
[68] Ibid. [69] Ibid. [70] Ibid.

dictum that 'every contact leaves a trace'. The task of the forensic scientist was to examine paint, fibres, bloodstains, soil, debris, cigarette ash, marks, and impressions—to emulate, in other words, the model laid down by Sherlock Holmes and so creditably imitated by Bernard Spilsbury. The mould set at the turn of the century sticks: forensic science is still almost exclusively an arm of the State. Its opponents are still dismissed as hired partisans whilst their Crown counterparts receive the law's uncritical endorsement. Impartial forensic science has come to mean State science. Writing in 1968, for example, Hamish Walls still described the most important function of scientific evidence as being 'to convert suspicion into a reasonable certainty of either guilt or innocence'.[71] Forensic scientists in the late twentieth century complain that it is part of Spilsbury's legacy that lawyers still believe that pathology 'is an absolute science, capable of absolute proof; that any expert worth his salt will go into the witness box, raise his hand and say "It is so, because I say so" and so demolish any counter argument, usually by force of character and rhetoric rather than by an assessment of the research'.[72] In later chapters we will explore the consequences of this legacy.

[71] H. J. Walls, *Expert Witness: My Thirty Years in Forensic Science* (John Long, London, 1972); H. J. Walls, *Forensic Science* (Sweet & Maxwell, London, 1968).

[72] Green, 'Is Sir Bernard Spilsbury Dead?'

6 The Law's Response

Throughout the nineteenth and twentieth centuries more and more experts were drafted in to give evidence on more and more topics. How did the legal system adjust to the influx of experts? One reaction was intense judicial activity designed to define further and more exactly the role of expert witnesses, to hedge their evidence round with rules and procedures. The need to structure expert evidence stemmed from the real possibility that experts would usurp the judicial role, and the fear that science would displace law as the touchstone of social order. This threat was becoming more apparent as attempts were made to render the law's own procedures and rules more scientific. The law's clientele was increasingly drawn from the prosperous middle classes, whose wealth depended in large part upon the application of science and technology. The law had to come to terms with the demands of its new clients for appropriate legal forms to express their new relationships and for improved procedures to handle them. The procedure and organization of the courts were deemed unsuitable for these purposes. They were said to be out of step with the needs of developing industry, commerce, and trade. In a sense, what we begin to see in the nineteenth century is a scientization of law. The Great Exhibition of London in 1851 had clearly established the connection between science and the prosperity of the nation.[1] The law was required to facilitate this connection:

> At a moment when the pecuniary enterprises of the kingdom were covering the world, when railways at home and steam upon the seas were creating everywhere new centres of industrial and commercial life, the common law courts of the realm seemed constantly occupied in the discussion of the merest legal conundrums which bore no relation to the merits of any controversies except those of pedants, and in the direction of a machinery that belonged already to the past'.[2]

[1] C. A. Russell, *Science and Social Change 1700–1900* (Macmillan, London, 1983), 236.

[2] Bowen, LJ, cited in W. A. Holdsworth, *A History of English Law*, vol. i (Methuen, London, 1956), 645.

The legal system thus experienced a direct push for progress along scientific and rational lines. It responded with a number of Commissions which were appointed to report on the courts and their procedures. These led to a number of recommendations being implemented in the Common Law Procedure Act 1854 and a series of Judicature Acts from the 1870s onwards. Simplified and rational codes of procedure were called for and new rules were promulgated. The jurisdictions of certain courts were amalgamated to make the administration of justice more simplified and rational. At the same time, a paid, uniform official staff was appointed according to regulations and duties laid down by an Act of 1879 (42, 43 Vict.) to replace an administrative bureaucracy riddled with patronage. This rationalizing of the system and codifying of procedure was said to make the legal machinery more 'suited to the needs of the present'.[3]

Many of the convoluted rules and procedures developed in this period sought to structure both the form and content of expert evidence so as to prevent experts from trespassing on the judicial province. This chapter deals primarily with the rise and demise of these rules and strategies. I deal here with four main rules. These are (1) the fact/opinion rule; (2) the ultimate issue rule; (3) the hearsay rule; and (4) the hypothetical question rule. However, rules such as these were only one part of the law's strategy of control. Other tactics included exploiting dissension amongst experts, portraying them as pedlars of uncertainty and a threat to order, discrediting them as the paid mouthpieces of advocacy, and, lastly, co-opting the best of them to support the established social and legal order.

Partisanship

The best explanation for dissenting experts that occurred to Victorian judges was that they must be suffering from partisanship. They spoke of experts testifying with great positiveness and in entire opposition to one another, of witnesses who

[3] Holdsworth, *History of English Law*, i. 649.

would rashly dogmatize and who were obscure. Some were said to promulgate frauds of justice and have contempt for the truth. It was said that experts could be found who would testify to anything absurd. Expert evidence was evidence of the lowest order and of the most unsatisfactory kind. Experts were hired advocates and their testimony was nothing more than studied argument in favour of the side for which they had been called. They were given to vain bubblings and oppositions; their aim was to mystify rather than to enlighten.

In 1873, Sir George Jessel, Master of the Rolls, decided that dissension was clearly caused by partisanship: 'Undoubtedly there is a natural bias to do something serviceable for those who employ you and adequately remunerate you. It is very natural, and it is so effectual that we constantly see persons, instead of considering themselves witnesses, rather considering themselves as the paid agents of the person who employs them.'[4] In 1876 Jessel developed his theory further. He referred to the practice of shopping around for experts.[5] According to Jessel, even though the experts themselves might be men of integrity, the manner in which expert evidence was obtained necessarily made their testimony unreliable. Experts might be men of integrity in their own habitat but they became venal wherever they came into contact with the law:

Now in the present instance, I have as usual, the evidence of experts on the one side and on the other, and as usual, the experts do not agree in their opinion. There is no reason why they should. As I have often explained since I have had the honour of a seat on this bench, the opinion of an expert may be honestly obtained and it may be quite different from the opinion of another expert, also honestly obtained. But the mode in which expert evidence is obtained is such as not to give the fair result of scientific opinion to the court. A man may go, and does sometimes, to half-a-dozen experts. I have known it in cases of valuation within my own experience at the Bar. He takes their honest opinions, he finds three in his favour and three against him; he says to the three in his favour, 'Will you be kind enough to give evidence?' and he pays the three against him their fee

[4] *Lord Abinger.* v. *Ashton* (1873) 17 LR Eq. 373, per Sir George Jessel, MR.
[5] *Thorn* v. *Worthing Skating Rink Co.* (1877) 6 Ch. D. 415, per Jessel, MR.

and leaves them alone; the other side does the same. It may not be three out of six, it may be three out of fifty. I was told in one case that they went to sixty eight people before they found one. I was told that by the solicitor in the cause. This is an extreme case, no doubt, but it may be done and therefore I always have the greatest possible distrust of scientific evidence of this kind, not only because it is universally contradictory and the mode of its selection makes it necessarily contradictory, but because I know the way in which it is obtained. I am sorry to say that the result is that the court does not get that assistance from the experts which, if they were unbiased and fairly chosen, it would have a right to expect.[6]

According to Jessel, consensus amongst scientists was to be expected. Payment for testimony introduced a mote into that otherwise perfect vision. More payment for more expert testimony introduced more motes. This proved a popular explanation amongst the Victorian judiciary. It made the idea of using independent court experts seem more attractive. Thus, by 1894, Justice Wills could blame the conflict he faced in *Kennard* v. *Ashman*[7] on partisanship. In this case, he solved his problems by adjourning the case to permit an independent surveyor to inspect the premises in question. In doing so, he found the consensus he wanted—instead of locking up his witnesses without food and water until they reached the required verdict, he simply replaced several experts by one expert. Clearly, even when the facts refused to speak for themselves with one voice, there were ways of making them seem to do so.

The view that disagreement amongst scientists was something abnormal produced by pecuniary interest had profound consequences for later expert witnesses. Recent sociology of science has demonstrated that disagreement amongst scientists is normal and natural. This being so, in order to meet the demands of Victorian courts for consensus, expert witnesses had to conceal their disagreements. Where they failed to do so they were lambasted as money-minded partisans. The legal system thus deterred any honest account of scientific work. Yet at one level, the law's attacks on the contaminated nature of individual experts' conclusions suggests that lawyers took

[6] Ibid. [7] *Kennard* v. *Ashman* (1894) 10 TLR 213.

seriously the claim that scientists were, after all, really disinterested seekers after truth. It was only the influence of outside interests which stopped them fulfilling this ideal:

> [One problem] with the notion that science is the disinterested search for truth is that it suggests that the importation of outside social interests will necessarily subvert the truth. The distinctions we make between pure and applied science or between science and technology hint that pure science is in some way thought to be 'better' than applied science, which is in turn 'better' than technology. Some kind of hierarchy based on the degree of purity of the activity seems to be a pervasive cultural theme.[8]

To Victorian judges it was nonsense for expert witnesses to claim to be real scientists at all. Their acceptance of the applied witness role made them impure and contaminated. The law's differentiation between pure and applied science allowed it to discriminate between what it was prepared to accept and what it was not. The distinction between 'pure' and 'contaminated' experts was further confirmed by experts themselves. As witnesses in an adversarial contest, they worked alongside lawyers to prepare and defend cases, to besmirch the reputation of a colleague, and to throw doubt upon his competency. Advocacy exploited the natural divides between scientists and further divided them from fellow professionals. It thus undercut ties of professionalism.

For lawyers throughout the nineteenth century, disagreements amongst men of science provided a resource for practical advocacy. Notions of impartiality and disinterestedness also provided an ideological resource for the rhetoric of law. This operated on two levels. On the first, the claim of science to exclusive and neutral knowledge of the truth made it a useful source of legitimacy for legal institutions under attack. On the other hand, disagreement amongst men of science provided, at the level of practice, good grounds for lauding the superior status of law as a means of dispute resolution. For courts of law, it became routine to dismiss expert evidence as the weakest kind of testimony, discredited evidence, unreliable, mere opin-

[8] J. Law and P. Lodge, *Science for Social Scientists* Macmillan, London, 1984), 134.

ion to be looked on with great suspicion. It was evidence which was readily procured by paying the market price. Advocates could laud the integrity of their own expert whilst accusing the opposition expert of being partisan and biased. Experts were witnesses whose evidence must be carefully scrutinized before it was acceptable. The truth must be sifted through the fine mesh of adversarial combat. Law, by providing this essential scrutiny, became society's safeguard against charlatan experts and their bogus claims. Dissent amongst experts was explained away in these terms, and the status of favoured (Crown) experts was maintained and protected by contrasting them with the dissident few. This line of argument also justified the judiciary's refusal to recognize experts as professionals and as scientists. In their view, it was clear that men who were prepared to sell their wares to the highest bidder lacked the altruism required of a professional. Mr Justice Neville complained that all expert witnesses were partisan, and that because they were men so lacking in integrity, they must not be allowed to invade the judicial province of the judge in the Patent Courts:

If the last twenty years were taken and an examination made of the voluminous shorthand notes which have accompanied the patent actions tried during that period, on their usual progress from the courts of first instance to the House of Lords, it would be found that a very large proportion of their contents, I should say at least nine-tenths, is devoted to questions which either openly or under a more or less skilful disguise are directed towards eliciting the opinion of the witness upon one or other of the issues in the case, or the construction of documents relied on . . . The amount of time wasted by this method of trying patent actions must be fabulous . . . but the waste of time is not the whole of the mischief, for the admission of the opinion of eminent experts upon the issues leads to the balancing of opinions and tends to shift the responsibility from the Bench or the jury to the witness box. The evil of this becomes aparent when one considers that, whereas expert witnesses called for the plaintiff almost invariably take a strong view in his favour on each and all of the issues . . . the expert witnesses for the defendants are equally confident the other way. It is rare to find any substantial difference of opinion between eminent experts upon matters of science whenever it is possible to dissociate the question from the immediate connexion with the issues in the action.[9]

The new scientists wished to gain a foothold in the law. Succeeding meant that they must appear to strike consensus. In doing so, they strengthened the belief that scientists would normally agree and that dissension was a product of partisanship. The needs of the law and the needs of the new expert professions thus precisely coalesced to produce a highly scientistic view of science. A new stick had been fashioned with which to beat the expert witness. The latter stood at the interface between science and the law. He stood also at a crossroads between the public and private faces of science.

The Fact/Opinion Rule

Having relegated experts to the witness box, the law began to hedge them in with new rules of evidence. These new rules required ordinary witnesses to separate fact from inference when giving their evidence. The rules prohibited witnesses from testifying to anything but the facts to which they were oyant or voyant, that is, they could testify only to that which they had heard or seen, and could not say what they thought, or believed, or had heard someone else say. This arose logically from the notion that it was the witness's role to supply the information but the jury's role to make deductions from it. However, courts soon found that 'Most language embodies inferences of some sort, it is not possible wholly to dissociate statements of opinion from statement of fact.'[10] The whole of our 'conscious life is a process of forming working beliefs or opinions from the evidence of our senses, few of them exactly accurate, most of them near enough for practical use, some of them seriously erroneous. Every assertion involves the expression of one or more of these opinions. A rule of evidence which called for exclusion of opinion in this broad sense would

[9] *Joseph Crosfield & Sons* v. *Technichemicals Lab. Ltd.* (1913) 29 TLR 370, per Neville, J.

[10] S. L. Phipson, *On Evidence*, 13th edn. M. N. Howard, R. May, and J. Huxley Buzzard (Sweet & Maxwell, London, 1982).

therefore make trials impossible.'[11] The rule made it difficult for ordinary witnesses to give their evidence in their own words; it forced them to break up the continuity of their story. Later judges were to complain:

> Every judge of experience in the trial of causes has again and again seen the whole story garbled, because of an insistence upon a form with which the witness cannot comply, since like most men, he is unaware of the extent to which inference enters into his perceptions. He is telling the facts in the only way he knows how, and the result of nagging and checking him is often to choke him altogether, which is, indeed, usually its purpose.[12]

Wigmore simply says that it did 'more than any other rule of procedure to reduce our litigation towards a state of legalised gambling'.[13] All statements of fact are, to some degree, statements of conclusion and judgement. Commentators on the law of evidence have recognized that 'in a sense, all testimony to matter of fact is opinion evidence, i.e. it is a conclusion formed from phenomena and mental impressions . . . Where shall the line be drawn? When does matter of fact become a matter of opinion?'[14]

In the immediate aftermath of *Bushell's Case* the courts did not in fact try to apply this rule to expert witnesses—they were asked for opinion as well as factual evidence and commonly gave it. The overall effect of developments in the seventeenth and eighteenth centuries was a structural shift in the expert's role from court expert and special juror to witness. This in effect reduced the expert's influence over legal decisions but it had the unexpected side-effect of cutting the court off from that element of his evidence it most required, that is, his opinion. The case of *Folkes* v. *Chadd* (1782) restored some of the expert's power by making him a special witness.

[11] McGuire, *Evidence, Common Sense and the Common Law* (Foundation Press, Chicago, 1947).

[12] *Central Railroad Co.* v. *Monahan* (1926) 11 F. 2d 212, per Judge Learned Hand.

[13] J. H. Wigmore, *On Evidence: A Treatise on the Anglo-American System in Trials at Common Law* (1940; Little, Brown & Co., Boston, 1983).

[14] J. B. Thayer, *Treatise on Evidence at Common Law* (Rothman Reprints, New York, 1969), 524.

The Ultimate Issue Rule

Making experts special witnesses meant that their opinions became authoritative and persuasive. A major worry for the judiciary was the prospect of experts trespassing on the province of the judge and jury. One of the main rules developed to prevent this was a rule prohibiting experts from commenting on the ultimate issue which the jury alone should decide. The deductions of experts were said to be based upon an assumed set of facts to which the jury was not privileged; these facts, in turn, were often based on the evidence of other witnesses so that, in effect, when the expert adopted one set of facts he was deciding which witnesses he believed. To do this was to usurp the role of the jury.

The precedent for the rule prohibiting experts from commenting on the ultimate issue in a case is usually cited as *R.* v. *Wright*, in 1821.[15] The substance of the rule is that no witness should be allowed to give his opinion directly on issues such as whether someone is guilty or innocent, sane or insane. In the nineteenth and twentieth centuries, judges have frequently reiterated this rule in order to prevent trial by jury from becoming trial by expert. The eminence of experts, combined with the special permission given to them in *Folkes* v. *Chadd* to hold forth in a manner denied ordinary witnesses, made them witnesses of great authority. Did a lay jury have the competence to evaluate the status of the expert, his claims to expertise, and the scientific details of his evidence? How could a lay jury reach a verdict which went against the learned views of the experts?

The law was caught in a double bind. Judges repeatedly defended lay juries as being capable of deciding complex issues, yet at the same time they were forced to concede that they required expert assistance. However, if the jury deferred too much to expert witnesses, this raised questions not only about the usefulness of the jury but about the utility of the legal forums as opposed, for example, to trial by experts. This

[15] *R.* v. *Wright* (1821) Russ. & Ry. 456, 168 ER 895.

threatened to undermine the tradition of trial by jury altogether. Juries might be hoodwinked by experts; they might be taken in by such things as appearance, or demeanour; they might be overawed by what the expert said and how he said it. Given the ready availability of expert opinion on the ultimate issue, the jury might be tempted to save itself the trouble of making up its own mind, for, 'if witnesses are too readily allowed to give their opinion concerning an ultimate issue, there is a serious danger that the jury will be unduly influenced'.[16] Cross cites the following as an example: 'If a cardinal of the Roman Catholic Church is testifying before a jury mainly composed of Catholics and states that, in his opinion, the defendant was driving negligently, it can hardly be supposed that the verdict would be other than for the plaintiff.'[17]

Keeping experts away from the ultimate issue was a central plank in this strategy to limit the expert's province. Yet if legal decisions were to secure some of the cultural authority of science they required the support of scientific witnesses. The rules about expert testimony should therefore not be so restrictive as to hamper experts from giving the courts the full benefit of their wisdom, but at the same time they should protect the judicial role. The issue was reiterated in 1913, in *Joseph Crosfield & Sons* v. *Technichemical Laboratories Ltd.*,[18] when Neville, J., stated that it was not competent for witnesses to express their opinion on the issues which the jury was to determine. In *Davie* v. *Edinburgh Magistrates*[19] the Court of Session went further than this in redressing the balance between the jury-box and the witness-box. It argued that neither the judge nor the jury was bound to adopt the views of the expert, even if they be uncontradicted, since 'The parties have invoked the decision of a judicial tribunal and not an oracular pronouncement by an expert.' Despite the existence of the rule and its frequent articulation in the courts, however, experts are asked questions on such issues and do give their answers.

[16] R. Cross, *On Evidence*, (4th edn., Butterworths, London, 1974), 382.
[17] 60 LQR 202. [18] *Joseph Crosfield & Sons* v. *Technichemicals Lab. Ltd.* (1913).
[19] *Davie* v. *Edinburgh Magistrates* (1953) SC 34 40.

Throughout the latter part of the nineteenth century the law strongly reiterated the doctrine that, whilst experts could say what in their view the facts were, and could draw inferences based upon their assumptions, they were not to comment on the ultimate issue, which was for the jury to decide. The ultimate issue rule meant that witnesses were not allowed to specify the physical cause of an act or condition, or the intent or purpose with which an act was done.[20] So, for example, expert opinion has been held to be inadmissible on such questions as why a deceased person committed suicide; whether someone voluntarily consented to a rape; whether a defendant could have inflicted a blow in a case of murder; whether a fire could have been extinguished; whether it would have been possible for someone to have fallen under a train; what started a fire; whether a deceased person killed himself; whether it was impossible that a wound had been self-inflicted; whether a defective piece of machinery was the cause of an injury; whether an article was obscene; whether or not the conduct of a fellow doctor was proper.

The Hearsay Rule

The doctrine, developed following *Bushell's Case*, that a witness must be oyant or voyant technically meant that the expert witness could refer neither to facts observed nor to opinions and theories developed by colleagues and predecessors. Nor could he refer to facts not in evidence before the court because these were not facts which had come under his personal observation. Anything he said about them was, technically, hearsay evidence and hence inadmissible. Thus, if in giving his opinion the expert drew upon the collected wisdom of his discipline, this was hearsay evidence and would normally be excluded.

In a sense, this rule was easier to apply in an age where there were few professional bodies and little accumulated professional knowledge. In the twentieth century, it has proved more difficult and impractical to apply for two principal rea-

[20] F. X. Busch, *Law and Tactics in Jury Trials*, 6 vols. (Bobbs-Merrill, Indianapolis, 1959–64).

sons: (1) the expert professions are now more numerous than they were when this rule was first fashioned; and (2) there has been a preservation and accumulation of knowledge within the professions, providing a storehouse upon which professional experts may draw, and within the confines of which they are generally educated. In addition, the expansion of the expert professions on an international basis, and the increasing specialization within disciplines, have both meant that new knowledge is constantly being added to the existing knowledge stock of a profession. The hearsay rule which excludes facts not personally observed attempts to operate in a world where knowledge has become more and more stratified and compendious. If it were ever possible for an expert to state all the facts upon which his opinion were based, in the modern world this aim has become highly problematic. As Freckleton says, 'No one professional can know from personal observation more than a minute fraction of the data which he or she must every day treat as working truths.'[21]

The effect of applying the hearsay rule to expert evidence was to make the giving of such evidence a long-winded and time-consuming business, as each of the facts upon which the witness based his opinion was adduced in evidence. It also had the effect of requiring experts to present their evidence in a highly artificial manner, contributing to greater reification of both the subject-matter itself and the practice of giving evidence. Strict application of this rule in the nineteenth century appears to have given way to a more relaxed attitude on the part of contemporary courts. The doctrine has been termed a technical prohibition rather than a principled objection. The law has increasingly allowed expert witnesses to include in their evidence facts and theories formulated by other people, which formed the underlying premiss or data upon which the expert's opinion was based.[22] In the case of *R.* v. *Abadom*[23] an

[21] I. R. Freckleton, *The Trial of the Expert* (Oxford University Press, Melbourne, 1987), 94.

[22] See *English Exporters (London)* v. *Eldonwall Ltd.* (1973) Ch. 415, per Megarry, J.; *H. and Another.* v. *Schering Chemicals Ltd. and Another* (1983) 1 WLR 143; *Seyfang* v. *Searle & Co.* (1973) QB 148; *R.* v. *Ahmed Din* (1962) CA, 46 Cr. App. R. 269.

[23] *R.* v. *Abadom* (1983) 1 WLR 126.

expert's reference to Home Office statistics on the refractive index of certain types of glass was contested on appeal as being hearsay evidence and therefore inadmissible. The Court of Appeal decided the expert had been entitled to make use of the statistics in forming his opinion. Reference to the statistical material did not make the statistics evidence and it was up to the court to weigh the probative value of that opinion. The test for expert hearsay evidence thus appears to have shifted away from issues of admissibility to issues of weight and probative value. In the case of *English Exporters (London) Ltd.* v. *Eldonwall Ltd.* (1973) Lord Megarry discussed the manner in which the expert opinion of a valuer was based upon the accumulated knowledge of the profession:

In building up his opinions about values, he will no doubt have learned much from transactions in which he himself has engaged, and of which he could give first-hand evidence. But he will also have learned much from many other sources, including much of which he could give no first-hand evidence. Text books, journals, reports of options and other dealings, and information obtained from his professional brethren and others, some related to particular transactions and some more general and indefinite, will all have contributed their share. Doubtless much or most of this will be accurate, though some will not; and even what is accurate so far as it goes may be incomplete, in that nothing may have been said of some special element that affects values. Nevertheless, the opinion that the expert expresses is none the worse because it is in part derived from the matters from which he could give no direct evidence [24]

This recognizes at least two things: (1) that expert opinions are typically a formulation of theories and experiments conducted either previously or contemporaneously by fellow experts; and (2) that the facts are theory-laden. Nevertheless, the law continued to require the distillation of facts. In 1974, in *R.* v. *Turner*, counsel was advised that he had always to ascertain from his expert witness what the facts were upon which he based his opinion:

Before the court can assess the value of an opinion it must know the facts upon which it has been based. If the expert has been misin-

[24] *English Exporters (London)* v. *Eldonwall Ltd.* (1973).

formed about the facts or has taken irrelevant facts into consideration or has omitted to consider relevant ones, the opinion is likely to be valueless. In our judgement, counsel calling an expert should in examination in chief ask his witness to state the facts on which his opinion is based. It is wrong to leave the other side to elicit the facts by cross examination.[25]

In the civil courts there has been some recognition that, even if the facts do speak for themselves, we do not always know what they say; indeed, we are not always aware of how much we know. At one level, the increased willingness of the courts to accept these difficulties represents an acknowledgement of the expert professions as equally erudite bodies of knowledge. But not all jurisdictions and not all professions have been accorded this accolade. In the United States, for example, experts are still likely to undergo an extensive *voir dire* on their expertise, and extensive examination on the factual basis of their evidence. Psychiatric and psychological witnesses in the United Kingdom are also still routinely required to lay out the entire factual basis of their opinion, and to engage in debates about the nature of the facts and opinions they adduce. The courts have, in other words, retained the right to challenge the accepted wisdom of the professions.

Psychiatrists have faced a particular problem where the facts they rely upon are deemed by the court to fall within the common stock of knowledge of a lay jury. The division between expert and lay knowledge is maintained by these disputes but it is nevertheless problematic. In certain areas of expertise, what was once uncommon knowledge sometimes becomes common knowledge. As Freckleton argues, what was once mysterious about some developments—for instance household electrical equipment, the working of the internal combustion engine, and so on—is now so commonly known that the lay jury does not require expert assistance to understand issues relating to such matters. With the growth of new technology in the twentieth century, new legal provisions have become necessary to permit the admission of documentary hearsay evidence

[25] *R.* v. *Turner* (1975) 60 Cr.App.R. 80, 82; also *R.* v. *Ahmed Din* (1962).

contained in such things as records and computer print-outs.[26] The Civil Evidence Act 1968 (Part 1) favours the admission into evidence of the opinion as well as the factual parts of an expert's report; the Civil Evidence Act 1972 and Rules 37 and 38 of Supreme Court Practice encouraged pre-trial disclosure of expert reports and 'do not accord any special relevance to the facts upon which the expert's opinion is based, whether they be hearsay or matters outside the expert's general professional knowledge and experience'.[27] Section 30 of the Criminal Justice Act 1988 makes expert reports admissible in criminal proceedings with the leave of the court and where it is not proposed to call the expert to give oral evidence. Case-law has encouraged the disclosure of scientific literature upon which the expert relied in making his report. At the same time, however, the courts have made it clear that, although the facts and literature relied upon may be vast and complex, they do not want experts to write a thesis.[28]

Contestable scientific evidence frequently becomes incontrovertible simply because it is never debated. An expert not called to give oral evidence cannot be cross-examined: his findings and opinions will not be tested. Moreover, it should be borne in mind that juries often find expert evidence compelling, and when that evidence is reduced to writing and introduced into the jury room, its authoritative status may be increased substantially.[29] Rules on hearsay thus contribute to the authority of fact.

The Hypothetical Question Rule

One main tool which the law devised to deal with the problems created by its fact/value, expert/lay judicial distinctions was a procedural device known as the Hypothetical Question.

[26] See Civil Evidence Act 1972; Police and Criminal Evidence Act 1984; Criminal Justice Act 1988 (which specifically permits documentary hearsay evidence when it would be 'impossible, impracticable or pointless to call the maker as witness'. See also P. Murphy, *A Practical Approach to Evidence* (Blackstone, London, 1980), 291.

[27] See RSC o. 37 and o. 38; for discussion, see T. Hodgkinson, *Expert Evidence: Law and Practice* (Sweet & Maxwell, London, 1990), 38–9.

[28] *Kirkup* v. *British Railways Engineering* (1983) 1 WLR 1165, per Lawton, J., 1170.

[29] Murphy, *A Practical Approach to Evidence*, 294.

This format was developed in response to a series of cases in which experts were thought to have usurped the role of the jury. Some judges allowed experts to sit in court, hear all the evidence, and then testify to their deductions. Their opinion was therefore formed on the basis of their observation of the other witnesses. In making their deductions they were making a number of untested assumptions about the truth of testimony given by these other witnesses. In doing this, experts were bypassing the jury's role in assessing the credibility and weight of testimony.

There were two central problems to be overcome. The first derived from the hearsay rule, which stipulated that experts could only give opinions based on facts presented in evidence before the courts. The second was that in giving an opinion the expert might be asked a question, the answer to which would either directly or indirectly imply a conclusion on the ultimate issue. In both instances, the judiciary constructed the problem in terms of the expert being a threat to the jury. In cases throughout the nineteenth century juries were repeatedly instructed by the judiciary that it was their task, and not the task of the expert, to try the accused. For example,

> What are the facts proved? Is Dr Sutherland to be the judge, or the Jury and I? It implies that the facts are proved; there are no facts proved at present. The way to examine a scientific witness is to put general questions to him such as 'What are the symptoms of insanity?' 'In what way do you judge such a symptom to be one of insanity or the reverse?' and the jury will apply his hypothetical answer to the facts before them; if it agrees with the facts, they will apply the opinion; if not, it goes for nothing.[30]

In *Rich* v. *Pierpoint* (1862), the problem was simply circumvented by using the hypothetical question.[31] Objection had been made to the evidence of a Dr Ramsbotham (Fellow of the Royal College of Physicians and author of a standard work on obstetric medicine). When asked whether he thought there had been any want of care on the part of the defendant, this

[30] *Ramdage* v. *Ryan* (1832) 9 Bing. 333; 2 Moo. & S. 421.
[31] *Rich* v. *Pierpoint* (1862) 3 F. & F. 35, 176 ER 17.

was said to be a question for the jury. The series of experts brought in after Dr Ramsbotham were asked the same question in the hypothetical form. This device effectively circumvented the objection that experts were infringing the jury's role whilst *de facto* allowing the experts to address the ultimate issue. This is what Hodgkinson means when he says that the courts have either tended to ignore the ultimate issue rule altogether, 'or to permit witnesses to give answers in careful words which appear to avoid being caught by the exclusion, though their true effect is to offend against the rule'.[32]

In the *M'Naghten Case* (1843)[33] one difficulty was precisely this problem of devising some means which would prevent experts from trespassing on the judicial province. The *M'Naghten Case* was fraught with political overtones, for

> M'Naghten's act took place against a backdrop of violence and profound national unrest. England was in a state of social, political and economic ferment. There were the Chartists and the Anti-Corn Law League, agitation for extending suffrage, radical demands for reform of child labour practices, and various schemes for experimentation in providing welfare relief. Radicals were rife and appeared to be threatening the structure of England. The temper of the times appeared to require the reassurance of definite and explicit rules. It was a time of violent feelings, and violent acts.[34]

In this atmosphere, the 'tragedies of insignificant people became severely overdetermined by social demands'.[35] Society, it was felt, deserved the full protection of the law which the novel insanity defence diminished. M'Naghten's trial came at a time when, as Smith writes, legislation had just set up a central Commission of Lunacy, along with which came 'career prospects and professional aspirations for a new medical specialism'.[36] The case required the law to deal with the claims of a new body of professional experts, the alienists (psychiatrists). The courts chose, instead, to rely upon the view expressed by

[32] *Shoemaker.* v. *Elmer* (1904) 70 NJL 710, 58A 940.
[33] *Daniel M'Naghten's Case* (1843) (HL) 10 Cl. & F. 200.
[34] J. M. Quen, 'An Historical View of the M'Naghten Trial', in C. R. Burns (ed.), *Legacies in Law and Medicine* (Science History Publications, New York, 1977), 247.
[35] R. Smith, *Trial by Medicine*, (Edinburgh University Press, Edinburgh, 1981) 174.
[36] Ibid. 3.

one Dr Isaac Ray, a medical man and author of *A Treatise on the Medical Jurisprudence of Insanity*, published in the 1830s. In their final determination of the issues the judges decided that a medical man conversant with the disease of insanity, who never saw the prisoner previously to the trial, but was present during the whole trial and the examination of the witnesses, could be asked his opinion as to the state of the prisoner's mind at the time of the crime; he could also be asked whether, in his opinion, the prisoner knew that his act was contrary to law or whether he was labouring under any delusion.

The medical men who had never seen the prisoner before the trial were Aston Key and Forbes Winslow. Both were eminent men in their field. They were allowed to add their opinions to the expert evidence of the defence. In the absence of rebuttal evidence from the Crown, the jury was charged to find for the defence a verdict of not guilty but insane. The case was reviewed by the House of Lords, which found the opinions of Key and Winslow proper. A medical man could give his opinion provided the facts in issue had been admitted and were not disputed. The issue was purely scientific. This conclusion was important. The scientific underpinnings of the legal verdict lent it a spurious air of objectivity, and hence freedom from political or social influences: 'Under the national stress of the time, the British judiciary redefined the common law of insanity so as to constrict it and to deny it the flexibility necessary for its adequate functioning'.[37] As Smith writes, alternative characterizations became the intellectual resources of groups with recognizably different interests.[38] Which views within a body of expertise the law accepts is a reflection of which side has succeeded in the competition for social acceptance; it is, moreover, also a determinant of which side succeeds. This frequently depends less upon the reason of argument than on the amount of material resources and backing it can muster. In this engagement, novel or unpopular

[37] Quen, 'An Historical View of the M'Naghten Trial', 247.

[38] R. Smith, 'Defining Murder and Madness', in R. A. Jones and H. Kucklick (eds.), *Current Perspectives on the History of the Social Sciences* (Jai Press, London, 1983), 175.

scientific theories, however sound they may be, stand at a disadvantage.

The hypothetical question proved to be popular strategy amongst lawyers. It circumvented the problem of asking an expert to give an opinion which inferred the validity of a disputed fact, and where the expert had no personal knowledge of the facts, these could be set forth in the form of a hypothesis to which he might respond. The hypothetical set of facts would parallel exactly the facts presented by the parties, that is, it was a mirror image of the facts in issue. For example, a man might be said to be suffering from XYZ symptoms. The lawyer would then say to the expert, *suppose* a man was suffering from XYZ symptoms, what would your opinion be? He might also ask, *if* he was suffering from ABC symptoms, what would your opinion be? Suppose, assume, and if are the operative words of the magic formula. Rather like a spell, the question would not work unless the proper words were uttered. The question and answer would technically become improper and inadmissible unless uttered in the correct form. If the jury ultimately found the facts stated in the hypothesis to be the true facts, the expert's deductions would be valid. A crucial advantage of using the hypothetical question was that questions had to be asked in certain ways and answers framed in the proper fashion. The expert was not to be allowed to 'roam through the evidence for himself, gather the facts as he may consider them to be proved, and state his conclusions on them'.[39] The hypothetical question was therefore a device which forced experts to fit their evidence into an artefact of the law's design. It was a lawyer's solution for a lawyer's problem. It gave lawyers extensive control over the form of an expert's testimony and consequently reduced the degree to which experts controlled their own evidence.

However, the initial attractions of the hypothetical question began to wear off with use. Every material fact in evidence had to be included in the hypothesis put to the expert. The trouble generated by this rule slowly became apparent.

[39] *Shoemaker.* v. *Elmer* (1904).

Lawyers included every material fact in their hypothetical questions: this included facts which were agreed as well as those which were disputed. This was soon found to be confusing and time-consuming. One hypothetical question, for example, ran to eighty-three pages, with fourteen pages of objections. By the time the jury came to its verdict it had at least two hypotheses before it, neither of which agreed, both of which had the full weight of expert authority behind them, and both of which had been obscured by counter-propositions put during cross-examination. Courts struggling with long and unwieldy questions decided that the solution was not to specify all the facts in the hypothetical question but only sufficient facts to justify the expert's opinion:

> While each hypothesis contained in the question should have some evidence to support it, it is not necessary that the question include a statement of all the evidence in the case. The statement may assume facts within the limits of the evidence, not unfairly assembled, upon which the opinion of the expert is required, and considerable latitude must be allowed in the choice of facts as to the basis upon which to frame a hypothetical question'.[40]

So long as there was no material exaggeration or perversion of facts assumed in a hypothetical question, technical accuracy in framing a question was not required.[41] The hypothetical question thus became a version of the facts in evidence rather than an accurate reproduction of the facts.

The detail with which the hypothetical question was framed also led to disputes. It could favour one side more than the other: 'An involved hypothetical question is ordinarily susceptible of more than one construction; hence words favourable to the party in whose behalf such question is propounded should not be stronger than those actually employed by the witness whose testimony forms the basis of a particular assertion from which a conclusion of fact is to be drawn.'[42] The disadvantages multiplied. It became possible to attack an expert for giving his opinion on an incomplete or distorted version of the

[40] *People* v. *Wilson* (1944) 25 Cal. 2d 341, 153, P. 2d 720.
[41] *Temple* v. *Continental Oil Co.* (1958) 182 Kan. 213, 320 P. 2d. 1039.
[42] *Guardian Life Ins. Co.* v. *Waters* (1943) 205 Ark. 87, 167 SW 2d 886.

facts. Crucially, it also became possible to undermine the expert's evidence completely simply by saying, 'You accepted the assumed facts as being true and your opinion is based upon that assumption?' followed by, 'You know nothing of course, of your own knowledge about the facts in this case?' or 'Your answer is merely your personal opinion on the facts recited to you?'[43] In cross-examination, the expert had to face counter-hypotheses and modified hypotheses, a set of substituted assumptions and questions designed to show that his opinion was not properly based on the assumed facts or that he had placed too much emphasis upon one of the assumed facts.[44] A whole array of cross-examination techniques could be brought to bear:

> A tactic resorted to by some trial lawyers is to ask the expert to state what he remembers of the hypothetical statement of facts upon which he gave his opinion. The hope obviously is that the witness will be embarrassed by not being able to remember some of the facts which are essential to his conclusion. Less experienced witnesses . . . will frequently answer . . . that in giving their opinions they considered some fact or factor outside the stated hypothesis; as for example, in the case of an expert medical witness that he also considered an attending physician's certificate or an autopsy report of a coroner's physician as to the cause of death.[45]

Another tactic was to ask the expert to consider a further assumed fact which challenged his opinion. His attempts to reconcile the two 'may discredit both him and his original opinion'.[46] Further tactics included requiring the expert to eliminate one or other of the facts assumed. He was then asked if this had altered his opinion: 'In such a situation the witness is driven to a choice between declaring that such facts have no determinative value or holding to his opinion with such facts withdrawn from consideration (which may make his previously stated opinion obviously absurd) or stating that he could not express an opinion with the suggested facts eliminated.'[47]

[43] Busch, *Law and Tactics in Jury Trials*, 139. [44] Ibid. 140. [45] Ibid. 141.
[46] Ibid. [47] Ibid. 142.

The variety of adversarial tactics encouraged by the use of hypothetical questions made it a minefield for experts. It had the reputation of being a complicated, long-winded, and confusing procedure. Learned Hand described it as the most horrific and grotesque wen on the fair face of justice. One court observed that 'asking a lengthy hypothetical question in a piecemeal method is objectionable if it is asked in such a way as to confuse the jury as to what the witness was, in the last analysis, supposed to answer'.[48] It clearly let in unimagined complications and led to all sorts of new questions about its form, propriety and utility. It generated multiple opportunities for procedural disputes and substantive questions. For example, what if an expert injected facts into the question out of his own knowledge, without telling the judge or jury that he had done so? 'The answer is rather misleading, and a hindrance rather than an aid to justice. It cannot always be known when an expert witness does a thing of this kind, but when it appears . . . from his own frank statement, the answer should be excluded at once'.[49] Introduced to ease the law's reception of expert evidence, the hypothetical technique thus actually made it much more complicated. Wigmore complained that it was misused by the clumsy and abused by the clever.[50] He argued that it led to an obstruction of the truth, that it artificially clamped the mouth of the expert witness, that it confused the jury, was a waste of time, and allowed counsel to build up a partisan account of his case.

Relaxing the Rules

By the beginning of the twentieth century, the judiciary's own control over proceedings had been consolidated by these rules on the form and content of expert evidence. The rules and regulations had more closely defined the expert's role and had established a set of conventions amongst lawyers aiming to produce expert evidence in court. It thus became possible to reduce the formality of the rules themselves as lawyers, judges,

[48] *Clautice* v. *Murphy* (1942) 180 Md. 558, 26 A. 2d 406.
[49] Wigmore, *Evidence*, para. 686. [50] Ibid.

and experts came more readily to rely upon established understandings of what was and was not permissible. What had once been objections to expert evidence on grounds of principle gave way to objections on the basis of technicalities. Instead of fighting experts law co-opted them and built up a cumulative set of conventions which regulated this relationship. New problems still arise, however. The law frequently changed its mind about what was an issue for the jury and what exactly was an issue for the expert, with the result that it is still not always clear when the expert will infringe the jury's role. In obscenity cases, for example, a series of high- profile trials in the 1950s, 1960s, and 1970s involved dozens of experts called to testify as to whether an article was likely to deprave or corrupt and whether it was obscene. In *R.* v. *Anderson and Others* some forty expert witnesses were called to testify that an article was not likely to deprave and corrupt.[51] The trial lasted for twenty-seven days, twenty of which were taken up with expert evidence from disc jockey John Peel, comedian Spike Milligan, and the Bishop of Woolwich, amongst others. The court stated that it was aware that 'some people, perhaps many people, will think a jury, unassisted by experts, a very unsatisfactory tribunal to decide such a matter. We can only deal with the law as it stands, and this is how it stands on this point.'[52] However, the convictions of the accused were quashed on appeal, the Court of Appeal finding evidence of bias against the experts called for the defence in the judge's summing up. He had called them 'so-called experts' and gave the impression that their evidence 'did not go for much'. The reasoning behind his thinking was that the lay jury did not require experts to help them decide what would be the reactions of the ordinary man.[53]

It was in the light of this case and others in a similar vein that the law decided that the layperson could equally well

[51] *R.* v. *Anderson and Others* (1972) 1 QB 304; (1971) 3 All ER 1152; see also *R.* v. *Calder Boyars Ltd.* (1969) 1 QB 151 CA; *DPP* v. *Jordan* (1977) AC 699, HL; *A.-G.'s Reference* (No. 3 of 1977) (1978) 3 All ER 1166, CA (Cr. D.).

[52] *R.* v. *Anderson and Others* (1972).

[53] *R.* v. *Chard* CA (1971) Cr. App. R. 268, 271; *DPP* v. *Jordan* (HL) (1977) AC 717, per Lord Wilberforce.

reach conclusions as to what was obscene.[54] Thereafter expert testimony as to whether an article was obscene was prohibited except where the intended audience was children, who could not be assumed to react in the same way as ordinary men.[55] It remains possible to call expert evidence as to whether the publication of an article is in the public good (the section 4 defence permitted under the Obscene Publications Act 1959) on the grounds that it is in the interests of science, literature, art, or learning, but expert opinion on whether the article is obscene is itself not admissible.[56] In 1985 an expert was allowed to give evidence as to whether an article dealing with methods of ingesting tended to deprave or corrupt since this was knowledge outside the realm of the ordinary jury, but the final decision on whether the article was likely to deprave or corrupt was a question of fact for the jury alone to decide.[57]

Views still vary as to whether, when expert evidence is given, the jury is under an obligation to accept it. On the one hand, expert evidence is not binding on the jury, on the other the evidence of experts is said to enjoy special weight and it would be a misdirection to the jury that it could simply ignore it: 'The tribunal of fact must obviously retain control over the findings of fact, which are its ultimate responsibility. This does not mean that expert evidence of a categorical nature, which is effectively unchallenged, may be disregarded capriciously in favour of unaided lay opinion, and it would equally be wrong to invite the jury to take this course.'[58]

The question of how far the expert should influence the outcome of jury trials has therefore never been fully resolved. It is hedged about and subject to the occasional bout of rhetoric but seldom is anything finally decided. This is a good example of the law keeping several options in the air at the same time. It is also probably true to say that this has been characteristic of the history of the relationship between the law and expert

[54] *R.* v. *Calder Boyars* (CA) (1969); (1968) 3 All ER 644.

[55] *DPP* v. *A. & BC Chewing Gum Ltd.* DC (1968) 1 QB 159; (1967) 2 All ER 504.

[56] *A.-G's Reference* (No. 3 of 1977) CA (1978) 1 WLR 1123.

[57] *R.* v. *Skirving* (CA) (1985) QB 819.

[58] *Anderson* v. *R.* (PC) (Court of Appeal of Jamaica) (1972) AC 100; see also Murphy, *A Practical Approach to Evidence*, 289.

testimony more generally. On the one hand the law insists that expert evidence which goes to the ultimate issue is inadmissible, whilst on the other hand courts every day admit it. Thus it is possible to find statements to the effect that 'the present trend of authority is to make no distinction between evidential and ultimate facts as subjects of expert opinion',[59] and that

We think the true rule is that admissibility depends upon the nature of the issue and the circumstances of the case, there being a large element of judicial discretion involved . . . Oftentimes an opinion may be received on a simple ultimate issue, even where it is the sole one, as for example where the issue is the value of an article, or the sanity of a person; because it cannot be further simplified and cannot be fully tried without hearing opinions from those in a better position to form them than the jury can be placed in.[60]

In practice, the rule that the expert must not comment upon the ultimate issue has given rise to so many problems that to all intents and purposes the law seems to have abandoned it. Thus Lord Parker commented in *DPP* v. *A. & BC Chewing Gum Company*,

I cannot help feeling that with the advance of science more and more inroads have been made into the old common law principles. Those who practise in the criminal courts every day see cases of experts being called on the question of diminished responsibility, although technically the final question, 'Do you think he was suffering from diminished responsibility?' is strictly inadmissible, it is allowed time and again without any objection.[61]

One commentator has observed that

it has not been decided whether the rule precluding expert opinion on an ultimate issue remains in effect at common law, though it has generally been assumed that experts need no longer be so confined. Certainly in civil cases it would seem to matter little whether or not an expert witness expresses in so many words to a judge who is the tribunal of fact what is obviously the necessary conclusion of his testimony on a relevant issue. In criminal cases, it may be that the trial

[59] *Millar.* v. *Tropical Gables Corp.* (Fla. App.) (1958) 99 S. 2d 589.
[60] *Hamilton* v. *United States* (1934) 73 F. 2d 357.
[61] *DPP* v. *A. & BC Chewing Gum Ltd.* (1968).

judge should retain the power to stop the expert short of doing the jury's work for them. This seems to have been the experience in the United States. For some time the Federal Rules of Evidence permitted experts to testify freely on ultimate issues both in civil and criminal cases. However because it appeared that such testimony might be accorded undue weight by juries in criminal cases, particularly those involving questions of the defendant's mental state, the relevant rule was modified to restore the common law position'.[62]

Cross argues that the tendency to ignore the rule is not limited to the English criminal courts, and cites an Australian case heard before the Privy Council in 1973 as an example. In that case, *Lowery* v. *R.*,[63] Lowery and King were convicted of killing a 15-year-old girl. They each admitted they were present but alleged that the other did the killing. The judge allowed Lowery to call a clinical psychiatrist to swear that King was immature and emotionally shallow whilst Lowery was a more aggressive personality. Cross notes that evidence of a very similar kind was introduced in a Canadian case,[64] but the Supreme Court of Canada had held it to be inadmissible because it came too near to answering the question which the jury had to decide. On Lowery's appeal to the Privy Council the trial judge was held to have been correct to have allowed such testimony. However, the Court of Appeal in 1975, discussing the case of *R.* v. *Turner*, sought to limit the damage done by *Lowery* v. *R.* by arguing that the decision in that case applied to 'special facts'. It should not, therefore, be taken as providing a general exception to the rule excluding expert opinion on the ultimate issue:

We do not consider that it is an authority for the proposition that in all cases psychologists and psychiatrists can be called to prove the probability of the accused's veracity. If any such rule was applied in our courts trial by psychiatrists would be likely to take the place of trial by jury and magistrates. We do not find that prospect attractive and the law does not at present provide for it.[65]

[62] Murphy, *A Practical Approach to Evidence*, 290.

[63] *Lowery* v. *R.* (PC) (Supreme Court of Victoria) (1974) AC 85; (1973) 3 All ER 662.

[64] *Lupien* v. *R.* (1970) 9 DLR (3d) 1.

[65] *R.* v. *Turner* (CA) (1975) QB 834; see also *R.* v. *Rimmer and Beech* (CA) (1983) Crim. LR 250.

The Court of Appeal has, from time to time, reiterated the notion that scientific experts are simply secondary assets in courts of law, that they have a subsidiary role which is to furnish information. Judges and juries are advised to beware of that old folk devil, the expert who dresses up his testimony in scientific jargon and has a long list of impressive qualifications. The full force of judicial rhetoric has been employed to underline the point that jurors are just as capable as experts of deciding matters of human nature. This grand defence of the jury has less to do with an ideological preference for the ordinary layperson, and more to do with professional power struggles between the judiciary and persons who have special knowledge. For this reason, if for no other, we should take what the judiciary says in its flights of lofty rhetoric with a pinch of salt. For other reasons too. Time and again, judges allow experts to give evidence on the very issue without demur. In *R.* v. *Mason*[66] the defence to a charge of murder was that the deceased had committed suicide. A doctor who had heard all the evidence was asked whether in his opinion the wound had been inflicted by someone other than the deceased. The Court of Appeal held that his answer was admissible because it was based on an assumed state of facts. Similarly, in *R.* v. *Holmes*,[67] the Court of Appeal decided that a doctor called in support of the defendant's plea of insanity could be asked in cross-examination if the conduct of the accused after the crime indicated that he knew the nature of his act and that it was wrong, even though, as Cross states, 'these are of course the very points which determine the applicability of the M'Naghten Rules'.[68] Lord Goddard, commenting in *R.* v. *Holmes*, said: 'Whatever fine distinctions may have been drawn in the days of the M'Naghten case, we can only say that no member of the court has ever heard an objection being taken to such questions as those. It seems to the court essentially a question that may be asked and

[66] *R.* v. *Mason* (CA) (1911) 76 JP 184; 28 TLR 120; 7 Cr. App. R. 67.
[67] *R.* v. *Holmes* (1953) 2 All ER 324; *Bleta* v. *R.* (1964) SC 561.
[68] Cross, *On Evidence*, 388.

answered.'[69] What was once a fundamental issue of principle is here downgraded to a matter of fine distinction.

Freckleton comments that courts of law are increasingly willing to recognize the necessity of allowing questions on the ultimate issue.[70] He gives examples which include cases where the issue was proper seamanship,[71] diminished responsibility,[72] and malpractice actions against professionals.[73] A doctor who has examined a motorist charged with drunken driving might also state whether in his opinion the accused was so drunk as not to have been in proper control of the car.[74] One rationale for allowing experts to testify in this manner is that their evidence is, after all, simply evidence: the jury is not bound by it and can reject it. This is another example of the judiciary keeping several options in the air at the same time: expert evidence, on this view, is simply evidence like any other, yet it is also said to carry great weight, and indeed the law itself makes experts special kinds of witnesses. If expert evidence were evidence like any other why has the law spent so much time and energy framing rules to limit its influence? Other writers have argued that expert witnesses should be allowed to express an opinion on the ultimate issue, subject to the proviso that the judge retains power to limit such testimony where there is a danger of the jury giving it undue weight as, for example, in cases involving defences of automatism, insanity, and diminished responsibility.[75] The Civil Evidence Act 1968 section 3 specifically provides that an expert may give an opinion on 'any relevant matter on which he is qualified to give expert evidence'. The old common-law rule prohibiting experts drawing conclusions on the ultimate issue was based precisely on the undesirability of allowing the expert to become involved in the decision-making process. However, the application of the

[69] *R.* v. *Holmes* (1953).

[70] Freckleton, *The Trial of the Expert*, 70.

[71] *Fenwick* v. *Bell* (1884) 1 Car. & Kir. 312, 174 ER.

[72] *R.* v. *Byrne* (1960) 3 All ER 1, 4.

[73] *Davy* v. *Morrison* (1931) 4 DLR 619, 626; see also *Fisher.* v. *R.* (1961) 130 CCC 1, 19–20, 24–5; *R.* v. *Searle* (1831) 1 Mood. & R. 75, 174 ER; *R.* v. *Rivett* (1950) 34 Cr. App. R. 87.

[74] *R.* v. *Davies* (1962) 3 All ER 97.

[75] Murphy, *A Practical Approach to Evidence*, 290.

rule sometimes amounts to nothing more than a play upon words.[76] For example,

In matters where technical details, complicated machinery or forces not understood by the ordinary man are involved, jurors may be assisted in reaching a just opinion of their own by learning the opinion of persons trained in special knowledge or experience of the facts concerned. To assist them is the function of the expert witness. If he is asked to express an opinion on the credibility of witnesses or to determine the issues involved his answer is to be excluded; but where an opinion is to be drawn from admitted or assumed facts as itself a fact to be weighed by the jury, his testimony is competent even though it may be the duty of the jury to form their opinion upon the same matter.[77]

Another view is that, since it is the province of the jury to hear all the evidence, including opinion evidence, to weigh it all, and to decide on the issues, an opinion cannot invade the province of the jury, even though the opinion is upon the very issue to be decided.[78]

Even in the nineteenth century, the old common-law rule was in trouble; in the twentieth century it appears to have been effectively abrogated. Murphy complains that the old rule produced great artificiality and that there is now no good reason 'why the expert should not be asked to deal with the point which everyone knows he is called to prove or disprove'.[79] There would appear to be little left of the old rule in the modern practice of law, especially since it is no longer applicable in certain civil cases. Despite the occasional bout of rhetoric from the Court of Appeal, the trend has been towards lax application of the rules and, in some jurisdictions, towards total abolition. Apart from the provisions of section 3(1) of the Civil Evidence Act 1968 in England and Wales, Rule 704(*a*) of the United States Federal Rules of evidence abolishes the ultimate issue rule (though in 1984 this was again altered in order to prohibit mental health professionals from testifying on the ulti-

[76] Cross, *On Evidence*, 389.
[77] *Coulombe* v. *Horne Coal Co.* (1931) 275 Mass. 226, 175 NE 631.
[78] *Eickmann* v. *St Louis PS* (1952) 363 Mo. 651, 252 SW 2d 122.
[79] Murphy, *A Practical Approach to Evidence*, 290.

mate issue). Freckleton notes that the Australian Law Reform Commission Research Paper no. 13 (1983) recommended substantial abolition of the ultimate issue rule.[80] The Criminal Law Revision Committee in its Eleventh Report made substantially the same recommendation.

Lax application of the ultimate issue rule, the opinion rule, and the hearsay rule all points to the flawed nature of the fact/opinion distinction more generally and to the problems of applying rules to enforce it. They were formulated at a time when the law courts were overwhelmed by the influx of new experts in the courts. In the twentieth century, they have declined to the point where there appear to be no real rules at all. Some judges thus accept expert evidence on the ultimate issue, others reject it; lawyers routinely ask experts to answer questions on the very issue and experts routinely answer them. Moreover whilst, technically, certain questions are open to objection unless framed in hypothetical form, it is up to the lawyers on the spot to take up the issue. Very often, they do not. The silence of the courts in such situations is permissive. It indicates that the policing of these rules is no longer a judicial priority. This amounts to an invitation to lawyers and experts to ignore them. Routinely they do exactly this. This is the way in which the rule of law normally goes on. When the courts adopt an active rather than a passive attitude, it is often more indicative of judicial insecurity than of any genuine concern about the expert usurping the jury's province. These are problems now more or less confined to the criminal courts, since the enactment of the Civil Evidence Act 1972, section 3(i), expressly provides for the admissibility of an expert's opinion on a matter in issue in civil proceedings. But since this rule remains on the books in the criminal courts, judges are entitled to insist upon its application whenever they deem fit, even though they are ill equipped to police its infringement. It would be a mistake to suppose that lax application of the rules means that the law has relinquished its attempt to control scientific experts. The hypothetical question format is still useful

[80] Freckleton, *The Trial of the Expert*, 78.

and still used. Although it provides a weak form of judicial control, it is judicial control none the less.

Does this more relaxed attitude mean that science and law have reached some kind of accommodation, or that science is now more able to resist the law's harness? Experts now have their own accrediting institutions and are therefore less dependent upon law as an avenue of social recognition. But though their power to resist the law's harness may have increased, the harness itself is still in place. There are still many rules which impose constraints upon experts and may deter certain kinds of experts from appearing as witnesses. This in itself indicates that the rules of evidence are not just empty rules in the books. Though they may not always be deployed, they form part of the legal repertoire. From time to time they are invoked, perhaps because the judiciary feels especially threatened, or because some tenacious advocate decides to object to a lax application of the rules. He generally does so because to insist on rule application in this particular instance will be to the material benefit of his client.

The structural placement of experts as witnesses means that they still remain vulnerable to accusations of partisanship. This provides the law with the classical strategy of divide and rule. It sets an expert to catch an expert. The law also retains a number of powers to which even the most eminent of experts may be subjected. It can insist that they prove their expertise in open court and it can, if it so desires, deny their province. In doing so, it can subject them to an unfamiliar and humiliating examination of their credit and status. It may require them to undergo challenges to their professional integrity and the validity of their views. It requires their professional deference and it can publicly chastise them when this is not forthcoming.

Having these rules on the books provides something more than the mere semblance of power. In appearance as well as in reality judicial control is exercised over (1) the verdict; (2) the style, form and content of expert evidence; and (3) the recognition of a man of science as a professional expert. The law still holds some sway over the making and breaking of professional reputations, as the following chapters will demonstrate. On occasion, its pronouncements have had a devastating effect upon the professional careers of experts who have appeared as witnesses. The fact is that the judiciary retains the power to confer or deny professional status, to humiliate and control expert witnesses, to reveal the interpretative uncertainties of their work, to shape the form and content of their evidence, and to insist upon their deference throughout the legal process. This acts as a significant brake on the conduct of experts in court. If they want recognition, they must abide by the rules of the club. One of these rules is that they must not speak out of turn.

There is a further reason why the rules in the books may now be more laxly applied than in the past. Behind-the-scenes controls on expert witnesses have increased and are now routinely exercised by the legal profession in the pre-trial stages of litigation. These control the market for and the supply of expert witnesses; they determine the form and shape of their evidence and their conduct in the legal process; they teach experts how to become skilled as witnesses; and they build up tacit understandings between lawyers and experts. Some of these mechanisms control witnesses generally; others are specifically applied to expert witnesses. The nature and effect of these controls are explored further in the following chapters.

7 Finding the Right Expert Witness

In the foregoing chapters I have argued that there has been a relaxation of the rules of evidence surrounding expert witnesses. I have also suggested that there has been a demise of grand judicial rhetoric on the subject of experts invading the judicial province. In its place, as I shall now try to show, the late modern legal system relies upon behind-the-scenes controls to locate and socialize experts. These backstage strategies ensure that good experts are selected, that bad experts are weeded out of the system, and that good experts become skilled witnesses.

Knowledge about experts forms part of a corpus of knowledge specific to the legal profession, usually to that part of the profession which deals with what is generically known as court work. Lawyers who do court work must know where to find an expert; they must also know who is and who is not an expert for their purposes. Once the need for an expert has been agreed, the task of finding the right one has to be accomplished. Some kinds of litigation, such as personal or industrial injury claims, use experts more frequently than others. Lawyers dealing with such work on a regular basis have routine strategies for finding and hiring the kind of experts they most frequently need, such as neurologists, orthopaedic surgeons, anaesthetists, psychologists, specialists in various kinds of limb fractures, post-traumatic syndrome, and so on. Other kinds of civil actions which routinely require expert input

The qualitative data used in this chapter and the following chapters is drawn from my fieldwork in courts and lawyers' offices, and interviews with litigation lawyers in ten London law firms, expert witnesses (forty), and barristers (ten). The initial fieldwork included working alongside lawyers in a City law firm for nine months, as well as shorter periods of observation in another eight firms. Most (but not all) of the 110 cases observed in this initial period were heard at the Royal Courts of Justice or the Central Criminal Court (the Old Bailey). Case files for a further 317 cases over a ten-year period were also analysed. Of the 110 cases observed in the initial period (1976-84) some later went to appeal either in Scotland or England in the following six years (1984-92); they were also observed at this level.

include environmental pollution, damage to property, contractual disputes, construction disputes, medical negligence, child custody proceedings, paternity actions, patent disputes, adulteration of foodstuffs, trading practices, and so on. Experts are probably used most frequently in personal injury litigation. They inspect the site of the incident, be it a road accident or an accident at work, assess the adequacy of safety procedures and the extent of injuries or damage sustained, and provide some prognosis about future effects and the recovery of the injured party. Paternity suits may involve tests for blood group or DNA profiling. In criminal cases, experts may be needed to provide scientific evidence of contact between an accused and the victim. This may also take the form of blood grouping and DNA profiling but also includes evidence of contact such as matching fibres, hairs, footprints and fingerprints, and so on Experts may also be called to testify to the way in which injuries could have been sustained or damage to property caused, to provide evidence concerning the content of suspected illegal substances, and so on.

Free-Lance Independent Experts

The bulk of the free-lance independent expert's work is high-volume civil cases. Unlike experts who work for the State in one of its forensic science laboratories, free-lance experts do not have a dependable supply of work all year round. This makes it difficult for them to pursue a career in forensic work. The main sources of work are loss adjusters, insurance companies, trade unions, and law firms. Since the work-load fluctuates income is less predictable. Most experts therefore gravitate towards regional or metropolitan centres where case turnover is highest. They also focus on areas of work in which there is a more promising flow of business, such as personal injury litigation and fire damage. Low levels of legal aid remuneration and the delays experienced in payment of legal aid fees mean that there are few independent experts engaged exclusively in criminal work. Some firms have been set up in recent years by former HOFSS scientists. For example, the private sector

Forensic Science Service Ltd. was set up in 1982 by Henry Bland, who had previously worked for the Home Office for sixteen years. In his evidence to the Home Affairs Committee, Mr Bland stated that his company had grown to be the largest forensic science service outside the government service. In 1988 it employed four full-time consultants, and had access to many others on an *ad hoc* and part-time basis.[1] Other examples include the aptly named Explaw, Hayward Associates, DNA Associates, General and Medical Legal Services Limited, Emmerson Associates, the UK Forensic Science Service Ltd., Network Security Management Ltd., R. W. Radley, Dr Giles Audrey, Taylor Associates, Varfiadis & McIntosh, and Robert Harris and Leslie Dick, a division of the Hambros Banking and Financial Services Ltd. Since no firm of free-lance experts can afford to specialize too narrowly, even those primarily interested in criminal work must supplement it with work from civil clients if they wish to be cost-effective.

In addition to the free-lance experts, university and National Health Service departments may, from time to time, supply the necessary expertise for defendants and plaintiffs. The Forensic Science Unit at Strathclyde University in Glasgow has, for example, proved a crucial resource for defence lawyers in criminal cases. However, facilities can never match those in the Home Office laboratories. No firm of free-lance experts can hope to match the facilities of the State, simply because the capital outlay required is too high. Equipment is specialized and expensive. Unless it is used on a regular basis costs will never be offset by the income it generates.

The absence of an institutional structure for free-lance experts means that they are a heterogeneous group. They lack a common professional training, common professional association, and common system of accreditation. They also lack a common professional voice. This dispersed nature of free-lance experts makes them difficult to locate. One issue for lawyers thus becomes how to find and identify a good expert witness. There are three main routes:

[1] First Report of the Home Affairs Committee, *The Forensic Science Service*, vol. ii: *Minutes of Evidence and Appendices* (HMSO, London, 1989), 167.

1. *Informal recommendation* (*a*) by client (relatively unusual); (*b*) by counsel (fairly common); (*c*) by other solicitors (fairly uncommon); and (*d*) by experts (more usual).
2. *Firm's own list of experts* compiled over years of practice and containing the names of experts already used and known to the firm. It will also contain remarks about experts' performance as witnesses.
3. *Talent spotting in court*, i.e. watching experts in action for the other side, evaluating their performance, and noting down their names.

As I shall show, these routes are actually endemic to the system—it is no accident that there is no institutional arrangement whereby lawyers can locate experts and vice versa. Attempts to form a data base to fill this gap have generally proved unsuccessful. This failure stems from the fact that lawyers look for something other than a list of names and qualifications to aid their choice of an expert. Moreover, knowledge about expert witnesses is in itself part of one's cutting edge as a lawyer. There are, then, good reasons why lawyers prefer their own informal means of finding the right expert. There is an essential connection between the ends and the means of the lawyer's search.

1. Informal recommendation

In his book *The Technique of Persuasion*, David Napley recommends that lawyers looking for expert witnesses should read the journal *Medicine, Society and the Law*.[2] They should also attend meetings of the Medico-Legal Society, where they might pick up information about potential experts. This is reminiscent of the Crimes Club in the latter half of the nineteenth century. The function is essentially the same: to meet in a congenial and gentlemanly atmosphere, and to discuss forensic science and medico-legal issues of the day with other interested men of law and men of science. What kind of information lawyers pick up here is not specified, though it will certainly include whether or not a given expert is socially

[2] D. Napley, *The Technique of Persuasion* (Sweet & Maxwell, London, 1975).

acceptable and well regarded by fellow lawyers and judges—briefly, whether or not he or she is acceptable to the legal establishment. The circle of experts and lawyers involved in such meetings is generally circumscribed by their proximity to London, where meetings are usually held. It is also determined by level of academic interest in forensic science and medico-legal matters. Although many lawyers live near enough to attend the meetings of the Medico-Legal Society, few choose to do so. Yet theoretically they should have a strong professional interest in attending—their work routinely calls for contacts with experts. Their principal reason for non-attendance seems to be that it would not be of any great benefit to them. They have their own ways of finding experts and these are much preferred.

Free-lance experts are generally far more enthusiastic about inter-professional communication than lawyers. For some the aim is to communicate to lawyers the nuances of the field; for others, the aim is to become known as experts in the market for legal work. This need to get one's face known is essentially a problem for experts, not for lawyers. Inter-professional seminars might provide them with a better knowledge of psychiatry but what they really want are experts with a knowledge of law. Finding this kind of expert means looking elsewhere. Personal recommendation by other lawyers ensures that experts have already passed through some form of positive vetting. For this reason, lawyers tend to stick to the informed advice of fellow legal professionals and experienced experts who are already well acquainted with their needs.

While the informal system of access suits lawyers, experts themselves find it less than satisfactory. They find it hard to gain entry into the expert witness market. This difficulty reflects the sponsored nature of entry into the legal élite, as well as the expert's place as a secondary player. Experts have no specific institutionalized place; they are provided by the parties not by the courts; their voice in legal proceedings is silent unless the lawyer chooses to invoke it. This in turn reflects the fact that the legal case is the lawyer's case: he hires an expert to perform a certain task. If the expert fails to do it satisfactorily, he will probably be sent a polite note of thanks with his fee and not be used again.

Since personal recommendation is the key to being selected, it is important to experts that their professional reputations have preceded them by word of mouth: 'We like to think that our name has become known through our work over many years.' These experienced experts are aware of the importance of personal connections in accessing legal work; they have successfully established a reputation amongst law firms as reliable expert witnesses. Once this link is formed, the lawyer–expert relationship is cemented over many years:

Most solicitors know us in this kind of work. If they don't they get our name from barristers and the like.

We get our work by recommendation between one and another.

I get my work through a solicitor who has heard of me.

Mostly I work for solicitors, most of whom have lists of experts in all fields, and information and recommendation **is** a matter of liaison between them.

One does not advertise; the work comes by virtue of one's reputation and getting known.

We are an established practice. We are regularly featured on the solicitors' own list, the Forensic Science Society list, The Shaw Society list, and we have an advertisement in the Bar list and Solicitors' Diary: the journals of the legal profession.

The expert is pinpointed by the solicitor out of his own knowledge or that of other professional practitioners he knows.

We were first instructed by a firm of solicitors in an action about 1907.

I became an expert witness by being invited in 1953.

In consequence of their long-standing partnership, lawyers and experts often knew each other well. Experts who competed in court were sometimes old friends:

Usually we all know each other; they are mostly good friends now.

More often than not I meet the same people—more and more so.

We meet each other regularly. There's certain 'plaintiff' doctors you get to recognize as such.

Choosing a known partisan is said to be counter-productive since it is important for experts to seem impartial. On the other hand, the system itself encourages an in-house attitude. Experienced experts develop personal friendships with the judiciary and the legal profession. One may, thus, expect a certain, conservative, homogeneity of outlook. Neophyte experts have to find ways of bridging the old boy network. Lawyers are also concerned that their standard routes do not prevent them from finding new talents. Informal routes allow lawyers to access the kind of experts they want but it also means that the pool of experts is small. This means that the same experts may be used time and again. Their slice of the market for expert services is safe. It also means that other experts must compete for what is left, though generally there is not enough work of a sufficiently lucrative nature to sustain more than a few experts in any one area of expertise. The pool of expertise will thus be very small indeed. Writing directly to law firms or advertising commercially is generally a less successful route. In order to obtain a regular supply of work one needs to be recommended; without regular court work the expert has no chance to build up a witness profile.

2. Firm's own list

Law firms keep lists of experts. Since such lists are a marketable commodity (being the prior advice of experts on expert witnesses) they are well-guarded secrets. As such, they are difficult to discover. Indeed, one needs some insider knowledge in order to spot their existence in the first place.

These lists are the most usual source of experts and can provide a fairly exhaustive dossier. Bad and good experts feature: it is equally important to know who not to use. Thus, the information penned in alongside an expert's name, brief though it may be, provides lawyers with guidance on the suitability of the expert. For example, one expert was described on a list as being 'a bit of an old woman in the witness-box'. Another expert was described as not being 'top weight . . . but extremely useful and thorough'. An engineer was described as 'extremely experienced . . . he can usually be relied upon to find an answer to problems which have defeated other experts'. By contrast, another engineer was 'not found to be particularly satisfactory; slow; needs continual chasing, not very thorough in answering questions'; he came with the added warning, 'Consult X before using.' A planning expert was described as being a 'very thorough and aggressive witness who specializes in giving evidence at major inquiries'. A marine engineer was dismissed as having 'a tendency to mumble and therefore not to be a convincing witness, but seems sound. Takes a hard line but does show considerable bias towards side instructing him, which can be counter-productive.' By contrast, a chemist was described as 'A professional expert witness, very experienced, especially in commodity disputes . . . advises numerous bodies including WHO, EEC etc. Very high powered.' Likewise another engineer was described as 'extremely competent and incisive and would make a good witness'. A planning expert had a 'very gentle manner; he is a good witness on major inquiries such as Green Belt inquiries involving intrusion into rural areas'. And a traffic engineer 'was recommended to us and proved very successful'. Another expert, a consultant analyst, 'produced a thorough report and was most helpful but since case not tried do not know how he would do as a witness—or indeed, whether he was right! Can probably be used with confidence.' One marine engineer was said to be 'an experienced man but his evidence lacks conviction which a more thorough preparation could have given it. Of the people in this field, the last to go before the red cards!' This last example comes from a large City firm which grouped, graded, and cross-referenced the experts on its files. Classification was organized around a traffic light principle. Experts on the green cards could be used without hesitation, whilst those on yellow cards merited cautious use. Experts on the red cards were risky. Hence the last remark against the name of this expert (who came bottom of the yellow cards). Another expert was described in the following terms: 'He's a very thorough one man band engineer. A very tough witness but tends to stray into other fields and hold forth on matters outside his instructions.'

The construction of such a catalogue is standard practice in the court departments of larger firms. It is an assessment of the expert's professional expertise together with his skills as a witness. Where separate departments within the legal firm all use experts, a central list forms an important part of the firm's corporate knowledge. Names of experts are canvassed and collated within the firm and sometimes from corporate clients with their own in-house knowledge of experts. Format is dictated by the size and needs of the firm. Hence smaller general practice firms tend to have one partner who deals with court work as part of his overall responsibilities. The list of good and bad experts is kept in his head rather than on file. Smaller firms have less need to centralize and formalize their lists. Provincial lawyers may personally know other professionals in the same geographical area, and mix in similar social circles. Lists within firms may also reflect a local bias or the nature of the legal practice: a firm in a coal-mining area may, for instance, have an extensive list of coal-mining experts but have no names of psychiatrists. Experts with local knowledge can be more useful in such circumstances than London-based professional experts.

Experts are generally unaware of the existence of these lists. Their names are obtained either through personal recommendation by other lawyers or by talent-spotting in court. Courts provide one of the most useful opportunities for lawyers to assess experts as witnesses. Lists require updating to include comments about current courtroom performance and caveats or hints for future use. A good list is a taxonomy of expert witnesses.

Local lists of experts can be supplemented by the Law Society's list. However, City lawyers find this option the source of much amusement—they 'wouldn't be seen dead' using the Law Society's list because 'half the experts on it are either dead or retired'.' 'In our case, we never use the Law Society's own list. I rather suspect that it would be of assistance in all those cases where one does not want assistance, but of very little assistance in those cases where one might want it.'

Unlike the firm's own list, the Law Society list is seldom

constructed from informal recommendation on the legal grapevine. It is more popular with provincial solicitors with little personal knowledge of the experts available in the field. Some experts are placed on the list as a result of their advertisements in the *Law Gazette* but experienced litigation lawyers shun self-publicists in favour of those recommended within the profession. The same applies to other experts who write directly to the Law Society asking for inclusion on its list. The Law Society agrees that 'Some of the experts on our list may be dead or retired, some may be old and therefore not so good in the witness-box and so on.' It nevertheless represents one means whereby experts can gain access to the legal system, though it promises not to be a particularly fruitful route. In addition, in the 1980s there were a number of attempts by commercial companies to establish an international index on which experts would pay to have their names listed. Law firms seeking an expert in their field would pay a fee to the agency for the supply of relevant experts. The matchmaking was never very successful and proved commercially unviable. Various professional societies have also compiled their own lists of consulting scientists, which they distribute free to interested lawyers. What these lists fail to provide is exactly that kind of information which lawyers want. It is information about how helpful experts are, what they are like in the witness-box, how well they prepare their reports. It is, in other words, the judgement of a fellow professional in the adversarial process. Moreover, whether a list is compiled for commercial reasons or by an independent agency makes no difference. The compilers of the Law Society list do not, for example, depend upon its accuracy for their livelihood; lawyers do.

3. *Talent-Spotting*

Lawyers watch experts working for rival firms with great interest. They observe them in the courtroom, seek counsel's appraisal of their performance, and ask their own expert for an evaluation. On the basis of this the name of the sought-after expert, and information about his or her skills as an expert witness, can be added to the lawyer's list without the expert ever knowing. The knowledge about who is expert, where he can be located, knowledge that there are any experts in the field at all, becomes part of the lawyer's professional know-how. Not only the content of the information but its mode of collation remains secret: thus there are problems for anyone wishing to access the system—expert, tiro lawyer, and researcher alike. Moreover, knowing that there are other experts in the field who might take different views allows the lawyer to shop around for the expert who will most favour his case. The rules of procedure do not prohibit this tactic. The adversarial system encourages it. By shopping around lawyers may discover a plethora of opinions; in the end only one will be selected for presentation. In a sense, then, the process of shopping around not only creates dissension, it also limits it, implying a greater background of consensus within the scientific community than may actually exist. Shopping around is a legitimate practice. The lawyer has a duty to the client to present the best possible case. If the expert's report is unfavourable, a lawyer can look for another report from a different expert, and can go to as many experts as he likes until he finds one who will say what he wants to hear. It might even be construed as negligent if he did not to look for a more favourable report: 'You do, of course, find solicitors who hawk their client round to a number of experts until they hear the opinion they want. 'Sometimes experts are slanters and highlighters, but a good expert will bring out the points favourable to your case. You might have to shop around for that sort of expert.' Expense limits the extent of shopping around and solicitors tend to pre-empt the need for shopping around by making sure their first choice can be relied upon. This is one reason why the accuracy of one's list is important: it saves time and money. The lawyer may stop the search because a number of experts agree with each other's conclusions, because time limits how long he can go on shopping around, or because money restricts how many experts he can ask to report. He might think again about how best to tackle his case, and choose the most favourable (or least damaging) of the reports he has obtained. The most persuasive of all reports—

and thus the most valuable of all adversarial tools—is the report which appears not to persuade at all.

Professional closure may sometimes force the lawyer to shop around. For example, in one case the lawyer was forced to shop around because the medical profession closed ranks: 'This is not a case of shopping around but of a clan closing ranks. I was unable to obtain independent advice. I was instructed in 1970 and thereafter I saw in excess of twenty experts. At the end of the day I have obtained the advice of a consultant psychiatrist.'[3] This suggests that there are limits to the market metaphor suggested by describing the search for expert witnesses as shopping around. The principal factors limiting an expert's availability are professional caution, monopoly of the market, and low or unstable remuneration. Wealthy corporate clients may also buy up all the available experts in the field, giving them a strong strategic advantage in any lawsuit.

Finding the Right Witness

A composite picture of the expert as member of an adversarial team emerges from looking at lawyers' ideas about good and bad experts. Images of the good and bad expert witness are cumulatively built up from the comments of the judiciary, as well as from the informal observations of other lawyers and other experts. In this fashion, informal reciprocal understandings accumulate. These conventions gradually obviate the need for more formal rules and guidelines on experts and their evidence. Expert witnesses absorb these conventions. Understandings become mutual: 'One becomes a good expert witness;' 'One isn't born an expert, one becomes an expert witness.' It is 'a trade learnt on the job'. Only those experts who are accepted as good experts will be employed on a regular basis. Those found wanting become *de facto* inexpert and will not be used again. There is therefore a fairly immediate incentive to find out what is required.

What is it that makes an expert a good witness? Comments

[3] *Parker* v. *Kent Area Health Authority* QB 1978 (fieldwork case).

on the lawyers' lists make it clear that they prefer people who are good experts and good witnesses: 'Obviously experienced experts make better expert witnesses than do inexperienced experts; arrogant experts do not make good witnesses at all!' Lawyers' preference for a practical rather than an academic expert stems in part from the difference between their styles of giving evidence. Hypercorrect speech hinders the impression of credibility.[4] It detracts from the authenticity of the evidence if witnesses try to mimic the kind of language they hear lawyers using in the courtroom. The comparison between academic and practical expertise is quite commonly relied upon in advocacy, as the following comments demonstrate:

There's a certain amount of hostility towards academics as the judge seems to think it's all theory and not like that in the real world. There is scepticism. They prefer a meticulous approach and practical help.

Courts prefer the professional expert—you don't get academics in our line usually.

I should think the courts prefer the professional expert.

An academic expert is of very little use at all; a professional, technical, and practical expert is better and it's best to have a bit of both. There are many cases where the judge doesn't understand the technical side and wants someone to confine the case down to some definite questions, to say what the issues are.

Sometimes the courts need academics but they don't like wafflers.

This is where you see the difference between the occasional expert and the professional expert witness. They know how to slant their evidence and can do—sometimes do—more to win a case than counsel. There's a way of pitching things at judges; you can't lecture judges. You should be short, give clear answers to all questions, so that the judge can write it down. It's bad if the other side has someone who knows what he's at and you haven't.

We normally instruct professional experts i.e. persons who make their living out of investigating and providing reports for legal proceedings and giving evidence in court. It is important that an expert should know how to give evidence in court and should understand how his

[4] W. O'Barr, *Linguistic Evidence in the Courtroom* (Academic Press, New York, 1982).

work fits into the overall legal strategy. For example, you do not want some disinterested academic professor who does not appreciate that it is important to win cases! We would prefer academics—until they open their mouths!

Really you want an expert to be both [an academic and a professional]. You want him to be highly qualified and you also want him to be able to write a lucid report, and you certainly want him to be a good witness. You can get some experts who are very good on paper but are hopeless in the witness-box. In some ways it is more important to have an expert who is good in the witness-box. In a lot of cases you will know the sort of evidence that he can give without his telling you that and all that you really want to do is extract it from him when he is in the witness-box. Certainly you would not instruct an expert who, when he got into the witness-box, went back on what he had said in his reports, and there are experts who do that.

We would select the one who we thought knew most about the subject and who we thought was going to be the best witness when he went into the box.

We would be put off if he was not a good witness in the witness-box, or if he gave us a confused, muddled report, or if it was quite clear that he did not really know what he was talking about.

Because it is to the fellow players that lawyers look for signs as to what is required in an expert witness, who becomes an expert is subject to substantial fluctuation: lawyers seek to interpret the hints and observations of the judiciary, the non-vocal signs of interest or lack of interest amongst the jury, and the casual off-the-cuff remarks of fellow lawyers and other experts. If the courts are thought to prefer professional experts, lawyers look for professional experts; if they cannot find one, they will reluctantly use an ivory tower expert who, with careful handling, can still make a good impression in the witness-box:

You choose an intellectual expert only if he is one who is going to make a good impression in the witness-box.

You go for the one whose belief is strong, not a 'smoothie'. But our side is one thing—theirs is another.

He must be an academic with a full command of the dispute and the literature but he must also be a good, precise, witness.

One lawyer commented in an internal memo: 'We did point out that it was one thing to be very highly qualified but quite another to give evidence in the witness-box.' Lawyers' initial choice of expert is made with this distinction in mind: 'It is our function as solicitors to decide what is relevant and how the case is to be conducted and to determine the overall strategy. The expert should really fit into that. We organize the questioning together. An expert who is familiar with a type of case and feels the solicitors are missing the point may express his views and try to guide them, but, of course, we work as a team and normally we each know our own particular function and there is no conflict.' One indication of whether an expert has become a good team member is the level of instruction he requires pre-trial. Time is said to be the great barrier to full coaching of experts. As a result, experts are shipped into the case at the last moment, given a hurried briefing, and kept in the dark about what they can and cannot do in court. Counsel rarely inform experts of the rules of evidence, they 'hope he knows already'. Experts are 'supposed to know them: it's part of the job.' They try to pick up the rules as they go along:

I basically know what the rules are: I've been doing it for twenty-three years.

No, I've never been told them. It's expected that I know them. Judges are usually flexible enough to allow us to put things straight.

The courts seem lax about the rules and realize they wouldn't get much expert evidence if they stuck to them.

They've never been explained to me but I know them roughly.

I've worked it out for myself: I wasn't told.

You learn the law as you go along, informally.

There are exceptions, however: 'I felt bewildered; I had no guidance at all. After a while you get to speak the language, but I have no idea about legal terms and wasn't sure if there were any rules. I've never been stopped from saying anything.'

Experienced experts accept that they may be required to look at the papers the night before trial; with luck, they may have a brief conference with counsel in the corridors of the courts. This lack of pre-trial consultation is indicative of the haste with which trial lawyers typically operate. The need for experts who can cope with the reality of last-minute consultation at the door of the court is a product of a kind of legal culture particular to trial courts. Lawyers themselves frequently read case papers only the night before trial. The pressure of the day of trial serves to concentrate their minds on the issues in dispute, whilst trial court organization itself tends towards last-minute changes of court time-tabling of cases, unanticipated alterations to strategy when witnesses fail to arrive, and so on. In such circumstances, lawyers need an expert who understands the vagaries of trial courts and who requires little in the way of pre-trial consultation. It is also sometimes a deliberate legal strategy to keep the pre-trial consultation brief. As we shall see in the *Confait* case, the legal strategy sometimes determines that experts should be kept in the dark so that they may sincerely believe 'that the definition of the situation they habitually project is the real reality. If a performance is to come off, the witnesses by and large must be able to believe that the performers are sincere. This is the structural place of sincerity in the drama of events.'[5]

The expert is rarely fully acquainted with the entire legal story; he has no power to insist upon a more extensive involvement with the case. The intermittent nature of his contact with the legal team may be seen as one means of distancing him from the case—distance being thought to lend objectivity—but it can also mean that his evidence is based upon incomplete information. This is true even for experienced experts who see themselves as para-legals, fraternizing in the precincts of the court, sitting beside the lawyer, advising on questions, tactics, angles, and so on. They do not need explicit coaching before trial:

We have experience of this on two occasions.

They used to try it, but don't very often now. Youngsters who don't know me try. Others realize there's no point. They do say, 'I shall

[5] E. Goffman, *The Presentation of Self in Everyday Life* (Penguin, London, 1969); see also E. Goffman, *Relations in Public* (Basic Books, New York, 1971).

not be asking you about that,' etc. I try to be pessimistic in my report so that my evidence in court will definitely live up to it.

How successfully experts resist coaching depends partly on their being able to convince the lawyer that they do not need it. In other words, they must convince him that they are already aware of the contingencies they may face in the witness-box and that they are more than capable of dealing with them. Experts may therefore do considerable work to make themselves into good witnesses without the need for explicit instructions from lawyers. In court they are switched on or off as the lawyer requires, and once the case is over the experts rarely receive any feedback. As bit players they are expected to have no compelling interest in the outcome of a case. Feedback comes from being hired again in future cases. This underlines the point that experts are subsidiary players in the legal process. Both pre-trial and during trial, it is the lawyers who control the process, in broad outline and in fine detail. As Bennett and Feldman have pointed out, the degree to which a lawyer can engineer the course of testimony

> depends a great deal upon the willingness of the witness to co-operate and his or her ability to respond to cues in the line of questioning. Some witnesses are more co-operative and responsive to cues than others. As a rule, expert witnesses . . . called by the prosecution are the most effective partner with whom to play out a tactic of co-operation. . . . [They] tend to be sympathetic to the goals of the state and they deliver a confident line of testimony . . . [They] get a lot of practice in trial situations. This hones their sensitivity to the tactical moves of their examiners. Not only does this make expert witnesses excellent players in the co-operation game, it also enables them to disrupt effectively the efforts of opposing lawyers to orchestrate their testimony.[6]

Some expert witnesses are more successful than others at resisting the attempts to get them to answer questions in Yes/No terms. In re-examination good experts pick up the cues from counsel which allow them to repair any damage

[6] W. L. Bennett and M. S. Feldman, *Reconstructing Reality in the Courtroom* (Tavistock Press, London, 1981), 124.

done to their testimony during cross-examination. (This was described as the art of 'coaxing a cat back from the end of the branch whereto it has fled'.) Faced with awkward witnesses, lawyers have recourse to a number of formal and informal means of control:

> Sometimes experts are reluctant to answer questions with a simple yes or no. They don't want to give their evidence in an unqualified way. This is where the edge creeps in: you try to get them to say something definite for your side or against the other.
>
> Experts try to put more in their answers than you require. They get over this with time and experience and we tell them we want clear answers on the issues.
>
> You can get round experts who are reluctant to answer yes or no by framing questions differently. There are some chronic qualifiers about. Counsel should hold the reins.

In fact, counsel do hold the reins, for if a witness refuses to co-operate he can be forced to do so: 'Normally a professional expert will not refuse to answer a yes or no question. He may point to the difficulties of answering it. If however he refuses simply because he knows that by doing so he will create a difficulty for his own side, I get the judge to order him to answer it.' In a very real sense, then, witnesses are compellable. Good experts, however, recognize the cues given by their own lawyers which invite them to elaborate upon their answers. They follow where counsel lead. In cross-examination and re-examination they exploit opportunities to slip in evidence previously left out. Good expert witnesses are perceptive enough to know where counsel wants them to stop but also sufficiently alert to make their own repairs. Some experts become too good at this. They attempt to introduce into evidence items not brought out by either side:

> It frequently happens that I'm not asked all the right questions. I get annoyed: the case might turn out quite differently, so I tend to wander, quite intentionally.
>
> I tend to give precise answers but I get round bad questions by wandering. Sometimes counsel has forgotten to ask me something, and I can often work it into cross-examination.

I don't usually have many problems in this direction because if counsel doesn't put the questions to me which will allow me to say all I want, I make the point to the judge and add a rider to another answer by way of illustration. If I get another opening, I keep on talking around the subject.

What lawyers see as the hallmarks of a bad witness may thus on occasion be attempts by experts to live up to the ideal that a court of law is interested in the truth, the whole truth, and nothing but the truth. They pose a threat if they volunteer information which the lawyer would rather keep out of evidence. However, such occasions are rare. Experts generally remain confused as to the exact nature of their duty in such situations. Do they have a duty to the court to reveal all they know, or do they have a duty to their own side to answer only those questions which they are asked—a point constantly reiterated by litigation lawyers each day in courts of law and backed up by virtually every trial lawyer's manual? The issue does not only arise during trial—it is unclear, for example, what the expert's duty is should he become aware, pre-trial, of information which is likely to be of benefit to the other side. Does he inform the other side's expert, or does he simply inform his own lawyers and rely upon them to judge whether or not the information should be disclosed? Where are the rules to guide his actions? Should he rely upon his own professional ethics and etiquette in what is, after all, quite a different setting? Or does he take the word of his legal mentors, who are the experts in the field of law? As we shall see, these have been crucial issues for expert witnesses in a number of cases.

Credibility and Incredibility Indices

The degree to which experts have been successfully trained as witnesses can be measured by looking at the extent to which they have understood what is required of them in their pre-trial and trial roles. Many experts are quite unaware of the extent to which their work is shaped by law and lawyers. They simply think that they are chosen because they have the right temperament for the job, without probing further about the exact nature of this temperament or how they come to acquire it. Experience is one means of learning strategies for courtroom success and survival:

You don't know what it is to be a witness until you have been a professional expert, and even then, you need practice.

Both being a good expert and a good expert witness are important—a good expert can tell a side the good and the bad points but he may not be very impressive in court. If you have the temperament for it, you'll be all right, but if you haven't got that, no matter how many times you go to court, you'll never get it.

Given that there is no organized profession of expert witnesses there is no alternative to learning on the job. There is no professional organization of forensic experts which lays down standards and ethics. Yet despite this, ideas about what an expert witness should be like have transmitted themselves in remarkably consistent fashion:

You must be a good expert first but it is important to be a good expert witness—you mustn't humiliate counsel or lose your temper, but remain placid and keep your mouth shut and your wits about you. This comes with experience.

The inexperienced expert brings along all sorts of paraphernalia to impress the court, but a theatrical performance comes across: you need to be more straightforward to impress.

As far as litigation is concerned, it's better to be a good expert witness than just simply an expert on the subject, because a good counsel can destroy a bad expert witness no matter how well informed. Usually an experienced expert will have found a happy balance in order for him or her to give sound evidence in an acceptable manner which cannot be destroyed.

It is more important for the parties to have a good expert witness. If they have a good expert but a bad witness, the court could get the wrong impression.

[Courts] look for the general presentation of evidence—the confidence they place in this relates to the way it is put over, the way it is done.

They like an expert who gives a clear and concise opinion in everyday language without being pedantic. They dislike one that lectures at length or gives long-winded answers.

The court is influenced by what kind of show the expert puts on in the box. If he makes a big omission it could be damaging, as could breaking down in cross-examination.

Barristers decide if I'm a good witness. They tell me whether to wear a suit, not to wear a red tie, and so on.

It's better to look 55 than 25 as the courts are influenced by your demeanour. I once knew a counsel say that the expert in the case had to be changed because, although his qualifications were excellent, he looked too young to be convincing.

They like psychiatrists less than anyone else as they are less definite than anyone else.

You obviously have to get the feel of the court and if you have a judge who does not have much time for psychiatric argy-bargy, you'll be told by your solicitor not to go into technicalities, keep your evidence simple and try to carry the judge with you.

Our experience is that courts react favourably to expert witnesses who are impartial and give their evidence well. They dislike experts who give patently biased evidence in favour of their clients and who do not give direct answers.

The legal profession is the primary source of ideas about what constitutes a good witness; it is also the source of cautionary tales about what happens to bad expert witnesses. Experts depend upon lawyers for their information about the case, the court, the legal process, details such as how to dress, how to speak, what to say, and where to sit. Since no one knows exactly how far judges and juries base their decisions upon legal reasoning and technical issues, credibility indices are actively sought as a guide to whether the judge liked the expert's evidence. The upshot of these signs of credibility is an ideal type of the good and the bad expert witness. O'Barr writes that witness impact is hindered by their being overly talkative, making too many qualifications, being slow or argumentative and therefore unconvincing, being too dramatic and therefore seeming phoney, and using unfamiliar jargon intended to make an impression, but which comes across as insincere.[7] Witnesses who use hypercorrect speech are stilted

[7] O'Barr, *Linguistic Evidence.*

and unconvincing witnesses. Some witnesses are too short-winded, others are too opinionated, antagonistic, and chronic qualifiers. The expert, if he is to be credible, must be none of these. He must have that air of unqualified certainty which comes of having a full grasp of the case, even though he may in fact have a very partial picture. As one barrister said, 'Very persuasive men may on occasion be preferred to one who is right.'

Professional experts do not let cross-examination deter them from becoming witnesses; their willingness to go back again marks them out. Some, however, are reluctant to participate in legal proceedings because of the dangers of cross-examination:

If they're shot down in flames the first times, they won't go back. Some learn, but most don't.

If it bothers you, you shouldn't be in the job; you'll never be any good. It has to be treated like a game, a battle of wits. I enjoy it, as long as I keep my nerve and don't lose my temper. If you do that, you'd best get out of the business.

It's good for the profession to be exposed to cross-examination. It gives you confidence in what you say.

Observation fails to deliver any concrete examples of the kinds of horror stories used to frighten experts. According to one counsel, 'These days it is *infra dig.* to grill an expert witness.' The horror stories of experts being shredded to pieces by the rapier thrusts of learned counsel are intended to structure interpretation and experience. Experts will always feel relieved that they have not met their worst nightmares. Having survived, they can think that this makes them good witnesses and their lawyers good lawyers. The system of word of mouth recommendation means that experts who fail to meet the requirements of advocacy are less likely to squeeze through the process of selection. The number of problem experts is therefore constantly reduced. This, in turn, means that lawyer–expert relations can proceed on the basis of maximum cognitive laziness. Friction between participants will be glossed over, tensions will be masked and managed. This is evidenced by the fact that expert witnesses so rarely experience any conflict between their roles as men of science and men of law. They have learnt not to speak out of turn (i.e. not to volunteer information on their own initiative). If they do experience a twinge of doubt, they have learnt that they should not voice it.

Conflicts in the Witness Role

Like most witnesses, experts attempt to resist the closed-question format of advocacy. Most prefer to hedge, qualify, and add riders to their answers, whereas counsel require simple, direct replies. This resistance is particularly acute where experts are being asked to comment upon the conduct or views of a fellow professional, when they are asked to reply in very definite terms:

Witness. Well if I say that was orthodox, you are making me say X's view was wrong.
Q. I am not making you say that.
A. I don't like the word orthodoxy. I can't answer that question if you put it to me in that form.
Q. I suggest to you that your opinion is and was the orthodox view.
A. Mr X differs from me on what should have been done. He would have assessed the situation and made his own judgment on that. Mr X's world is a world of which I know nothing . . . It's an idiosyncratic ivory tower view.
Q. But in your opinion it was a reasonable thing to have done?
A. Everyone wouldn't have done the same but I can see why he did it . . . It was a reasonable view for him to take.
Q. I put it to you that he made a judgment no competent practitioner could have made.
A. I don't know that that's a fair or balanced judgment. I think it is too hard . . . One can't say a colleague was wrong just because one wouldn't do it oneself.[8]

Lawyers on either side seek to exploit or to diminish conflicts between the experts as it suits their case. Although these disputes arise out of individual claims, their broader basis is the law's policing of the professions and its insistence that professional judgement be brought within the limits of legal

[8] *Parker* v. *Kent Area Health Authority*.

categories. The law uses fellow experts to assist it in this task. The practice was set out in *Ramdage* v. *Ryan* in 1832: 'Because a jury must be ignorant of the conventional rules of etiquette established in each profession, which can only be known or only accurately known by members of the profession so that it is only from the estimation of his brethren that the public can judge whether an individual conducts himself uprightly in those matters with which he is most concerned.'[9] The problem which arises out of this principle of allowing experts to give their opinion about fellow experts is that it tends to allow experts to decide the very issue which is for the judge and jury to decide. As we have seen, great efforts were made by nineteenth-century judges to ensure that experts were not allowed to do this. In a number of cases, the matter was said not to be one for expert opinion at all, but for common sense:

> It is not a matter for the doctors. They may have their own views about it, but that is not better evidence than any other. It is not a medical question.[10]

> Some people, perhaps will think a jury, unassisted by experts, a very unsatisfactory tribunal to decide such a matter. We can only deal with the law as it stands, and this is how it stands on this point.[11]

The expert is a professional colleague but also a courtroom rival. Individual experts are expected to manage this tension. Since they are usually reluctant to criticize their colleagues in public, it is generally more difficult to locate an expert who is prepared to stand up in court and criticize the work of a fellow expert. For example, in one case the expert was not prepared to say that the actions of another doctor were negligent:

Q. There's a difference between us which could merely be linguistic.
A. All I am saying is that a reasonably competent man can sometimes make unreasonable judgements.
Q. It's not a decision a competent man could take?
A. There's a linguistic difficulty between us. Competent and capable people don't always act as such. It doesn't seem unreasonable to me.

[9] *Ramdage* v. *Ryan* (1832) 9 Bing. 333, per Tindal, CJ.
[10] *R.* v. *Ahmed Din* (1962) CA, 46 Cr. App. R. 269.
[11] *R.* v. *Anderson and Others* (1971) 3 All ER 1152.

Q. And not a decision that could have been taken by a reasonably competent man.

A. My words have been twisted. I said before what I thought. There are different opinions. That's why I'm here.

Q. And because there are different opinions, you can't say that it was an unreasonable decision. You will jump from on extreme to the other, Mr N!

A. It's just my view.

Q. But since the plaintiff is relying upon it, I have a right to test it.

A. I won't say that I'm not sure. I don't think anyone should make a categorical statement.

There is also a tension between delivering the kind of categorical opinion which the lawyer seeks from experts and their own professional survival in the witness-box: 'I can't state it with certainty: I might get caught out by counsel if I do so.' Some counsel lure the expert on further than he need go and then 'clobber him on it'. Some 'make a complete hash of it', others do it with great economy:

> I have seen doctors cross-examined on the basis that they are prejudiced against the plaintiff and I once saw an expert accused of giving false evidence. Obviously there are different ways. Each counsel has his own tactics. But if you can't get an expert to alter his views, you may be able to get him to temper them. You might say, 'You have expressed an opinion but of course you are not going to say that it is the only opinion that can be held, are you?' or, 'Presumably you accept the opinion my doctor holds as being held by some of the most eminent doctors in the land.'

One way of avoiding admitting the validity of the other side's opinions is to conceal any doubts one has about the certainty of one's own judgement, 'After all, I only get one bite at the cherry. If I do have any doubts, I keep them to myself until I can tell counsel privately.'

Training Better Witnesses

Training, as much as selection, reconstitutes ideas about the good expert witness. In an article written in 1988, Professor David Gee commented on the need for training amongst expert witnesses. He 'began to suspect that there was much more planning than I had thought behind the questions I had been answering all these years'.[12] The possible pitfalls have led experts to seek instruction in forensic skills. Several bodies have now produced introductory texts, organized workshops, and provided expert witness training courses. What kinds of skills are they teaching? Training tends to cover the preparation of reports, provision of samples and styles for future use, an outline of the stages of legal proceedings, explanation of legal terms such as writ, plaintiff, defendant, submission, balance of probabilities, pleadings, discovery, and so on. Sometimes terms such as relevancy and materiality are explained but it is doubtful whether experts can understand their full import without further exposure to concepts such as probative weight and the structure of proof. Gee argued that this kind of training would make experts better witnesses, equipping them with the necessary tools with which to survive in court. However, most advice to experts is not about their survival. It is about the survival of a legal case. One text for surveyors advises that every case has at least one strong point and one weak point, which the expert should help identify, reveal, and deal with.[13] It also advises that 'for a valuer to conceal something which he knows to be relevant would be ethically wrong; he is bound by his oath to tell the whole truth as he sees it'.[14] But of course the oath only comes into play once the expert has stepped into the witness-box—there is no equivalent undertaking in the pre-trial process. If the whole truth is to be made known to anyone it is to be made known only to one's own side.

Advice to experts from within the professions mimics lawyers' priorities. One writer advises experts, 'Only answer what counsel asks. Do not volunteer additional information unless clearly invited.'[15] He goes on, 'Under cross-examination

[12] D. Gee, 'Training Expert Witnesses', *Medicine Science Law*, 28/2 (1988), 96.

[13] J. A. F. Watson, *Nothing but the Truth: Expert Evidence in Principle and Practice* (Estates Gazette, London, 1975), 27.

[14] Ibid. 30.

[15] P. S. D. Reynolds and M. P. King, *The Expert Witness and his Evidence* (BSP Professional Books, Oxford, 1988), 119.

the expert cannot be compelled to disclose privileged information, advice or communications between himself, counsel and solicitors.'[16] Moreover, the expert should 'Be careful to answer the precise question and nothing else. He should make his answer clear and unambiguous.'[17] Watson's text for surveyors states:

> what the witness says will lose conviction if facts and opinions seem to be dragged out of him. He should always discuss with counsel beforehand the manner in which counsel will take him through his proof. . . . be careful not to introduce irrelevancies and try to deal with the various items of your evidence in the same order as they appear in your proof; counsel will tick them off in the margin of his copy. Personally, when about to give expert evidence, I used to suggest to my counsel that as far as possible he should confine himself to general questions: 'Can you assist the tribunal further, Mr Watson, by developing that statement?' Then I asked him to give me my head until I paused for the next question. Naturally I said that were I to forget something I should rely on his putting a question to remind me.[18]

Experts are warned that in cross-examination counsel will try to ridicule them, confuse them, and generally attempt to get them to change their views. 'However an expert must not be shaken and must hold to his view. If he changes his view it may have drastic consequences.'[19] There is, then, an understanding that convincing stories are strong stories firmly adhered to—contingency, fragility, and weakness are to be avoided. Clearly, it is not enough for lawyers that an expert holds an honest opinion: he must also be a good witness. He must find ways of conveying his evidence in a convincing manner. Watson advises that the most convincing expert witness is courteous and unflappable.[20] He should keep his evidence short, be willing to admit to a crashing mistake and to concede a minor error, stick to his area of expertise, and firmly decline an invitation by opposing counsel to express an opinion on matters outside his field of expertise. Further, the expert witness

[16] *The Expert Witness and his Evidence* (BSP Professional Books, Oxford, 1988), 118.
[17] Ibid. 115. [18] Watson, *Nothing but the Truth.* [19] Ibid. 76.
[20] Ibid. 83.

should resist any temptation to score off his adversary, avoid acquiring mannerisms, give his evidence in short words, avoid clichés, shun genteelisms, and abjure jargon.[21] They are told that judges do not like to be lectured, that it is up to the court to decide matters, and that they should give their evidence clearly, concisely, and preferably slowly, so that the judge has time to write it all down. They should keep an open and enquiring mind whilst bearing in mind the legal implications of the facts, bring these to the attention of the lawyer, and know what facts will support a legal argument and what will not.[22]

This is all advice on the art of impression management. It takes the priorities of the legal profession for granted and seeks to pass them on to would-be experts. In this way, the legal profession is able to build up professional loyalties and stabilize experts' expectations of their role. Becoming an expert witness involves what Barnes has called the subprogrammes of a speciality which, 'when added to a pre-existing pattern of socialisation, produce the capability for a certain kind of understanding and activity'.[23] An expert is rarely advised that he has a duty to himself not to utter an opinion he cannot justify or prove. Failure to abide by this rule could 'harm the client's case, the expert's credibility and his career'.[24] Members of the Association of Consulting Scientists are also reminded that they have a duty to their profession not to supplant one another. Yet helping counsel frame questions for the cross-examination of the other side's expert will be one of their main tasks as an expert witness:

> You can help your client's counsel enormously by telling him the answer to a possible attack, thus supplying him with material for cross-examination. To be forewarned is to be forearmed.[25]
>
> Once you have finished your evidence you may resume your role of assisting the legal team. Part of your duty will be to assist leading and junior counsel at the trial and to be present when the other side's expert is giving evidence.[26]

[21] Ibid. 118. [22] Ibid.
[23] B. Barnes, *Scientific Knowledge and Sociological Theory* (Routledge, London, 1977).
[24] Watson, *Nothing but the Truth*, 33.
[25] Ibid. [26] Reynolds and King, *The Expert Witness*, 119.

Despite the existence of witness-skills training therefore there remain deep conflicts in the expert's role. These are further exacerbated by structural inequalities in the provision of forensic expertise which determine that, for experts to make a career, they must work almost exclusively for one side in the litigation process.

Inequality of Resources: Capital versus Labour

For experts to have a regular turnover of forensic work which is sufficient to sustain a career they must work either for large corporations as in-house experts or for the State. The position of these experts is somewhat different from that of free-lance experts, who typically work for individual clients on legal aid. In addition, only some areas of scientific and technological work can sustain a large corpus of experts. In less profitable areas, there will be fewer experts. Personal injury and loss adjustment cases are profitable areas. They supply the bread and butter of the free-lance expert's work. However, relatively few experts are willing to undertake legally aided work. Those who do are seldom the top experts in their field. A study by Dingwall and Felstiner on asbestosis litigation confirmed that there were only about one or two dozen specialists in the whole of the UK.[27] There are a number of reasons for this. One is professional disdain—some experts simply do not wish to work for poor people nor do they wish to become involved with the type of problems generally associated with poor people. This may in part be a reflection of lawyers' own preferences—some law firms, for example, have separate waiting rooms for their private and legally aided clients so that the two types of client will not meet. Another reason is the wish to avoid conflict with professional colleagues, especially if they work within the same institution.

Corporate clients and government institutions tend to have their own in-house pool of experts employed on a full-time basis. The same is true of various manufacturing industries,

[27] W. L. F. Felstiner and R. Dingwall, 'Asbestosis Litigation in the UK: An Interim Report' (Centre for Socio-Legal Studies, Oxford, 1988).

construction companies, mining companies, and so on. The nuclear industry, for example, has its own medical experts as well as experts in nuclear science. It is often difficult to gainsay the evidence of these in-house experts partly because it is difficult to find an expert outside this circle who is willing and qualified to act for a plaintiff, and partly because it is difficult to gain access to in-house medical and safety reports. Internal reports are sometimes written with the possibility of litigation in mind, with the result that some in-house experts edit their reports in case they are ever produced in legal proceedings.

The tactics of some trade union law firms sometimes exacerbate the secrecy surrounding claims and settlements for industrial injuries. They generally prefer settlement to costly contest. In one trade union firm, for example, four asbestosis claims went to a negotiated settlement not because they would not stand up in court, but because the union considered it was not cost-effective to go to trial. Trade union firms may withdraw funding from plaintiffs who refuse to accept a negotiated settlement. According to the Pearson Commission the plaintiff's case against the employer is tested in approximately only 1.5 per cent of cases, and research more generally has shown that about 90 per cent of civil cases end in settlement. Most expert evidence is therefore never tested in court; employer liability is never established. This is, of course, the main advantage to employer companies, since no precedent is established upon which future claims may rely. As Green has pointed out, both trade unions and insurance companies may have long-term interests which override the short-term gains to be achieved in any given case.[28] Greater pre-trial disclosure, particularly of medical evidence, was recommended by the Law Reform Commission's Committee on Personal Injury Litigation in 1969. In the next decade, attempts to accelerate civil litigation supported this idea but:

It was felt that early disclosure, particularly of medical reports, could force plaintiffs to trial before the future course of their disease had become clear, or denying the start of a suit until the clinical signs

[28] J. Green, 'Industrial Ill Health, Expertise and the Law', in B. Wynne and R. Smith (eds.), *Expert Evidence* (Routledge, London, 1989).

were obvious, which would risk limitation problems and might, paradoxically, delay compensation even further . . . Some of these solicitors were also concerned that early disclosure would disadvantage their clients in pre-trial negotiations. . . . The art of representing plaintiffs lay in the ability to conceal from the defendants the real strength of one's hand. One admitted response was to encourage doctors to write much vaguer medical reports for disclosure in all cases and to make their real assessments in confidential letters, with the intention of making the defendants guess at the weight of evidence. This seems likely to make defendants even more reluctant to rely on the medical evidence lodged by plaintiffs and to seek more rigorous investigations on their own behalf.[29]

By far and away the bulk of the expert's work lies in this kind of out-of-court advocacy at the pre-trial stage of case construction. The absence of a court hearing with its attendant publicity is an advantage to the employer. Plaintiffs making a claim without knowing about prior accidents and illnesses are in effect rehearsing situations which others have already gone through. It is more difficult for plaintiff experts to discover this kind of information, whereas defendant experts typically have greater access to items such as safety records. However, since they are working for the defendant, they are bound by adversarial norms not to reveal this information to the other side. Important information about similar incidents may thus be withheld from the plaintiff's solicitors. Moreover, cases which do proceed to court come to be seen as deviations from the norm, rather than the norm itself. The net effect is the formation of a normal background understanding of work-place practices:

Suppose we cite metal fatigue as the cause of an air crash. The implication is that in other relevant aspects, the aeroplane was normal, how we would expect it to be. Had the metal been likewise, in its normal condition, no crash would have occurred. The deviation from normality explains another deviation from normality. The background of normality forms the act of unchanged necessary conditions against which the causal story stands out and operates as an intelligible communication. The cause is a necessary condition in which we

[29] Felstiner and Dingwall, 'Asbestosis Litigation', 11.

are interested; in labelling it we define a taken for granted background of normality. In another context, we might have said that taking off in a plane with metal fatigue caused the crash. The implication would have been that such aircraft were not normally flown.[30]

The prevailing background of normality informs the climate of industrial injury claims and levels of compensation. Experts may well have access to information which would help the plaintiff's case but the norms of advocacy determine that this is not disclosed.

The difficulties faced by plaintiffs are compounded by the low level of remuneration for experts working on legal aid. Working for corporate clients also promises a regular rather than intermittent flow of work and enables the expert to build up a long-standing relationship with the client. Corporate clients, including some large insurance firms, also have access to better facilities and resources with which to out-gun plaintiff experts.

Inequality of Resources: State Versus Accused

There are few large firms of free-lance experts able and willing to carry out work for the defence. Relatively little is known about the free-lance specialists in criminal work, apart from their heavy dependency upon criminal legal aid and the fact that this puts them at a considerable material disadvantage in relation to the Crown. Some light was thrown on the situation by Henry Bland in his evidence to the Home Affairs Committee. Having worked as a forensic scientist on both sides of the criminal justice system, he was able to compare his experience of the degree to which financial restrictions limited the work done by the Home Office Forensic Science Service (HOFSS) and by the defence. As an HOFSS scientist he was never restricted from making inquiries on the basis of cost. As a free-lance expert his position was quite different:

We had a murder and we asked for £900 to do a scene examination, night photography and it was certainly going to mean a

[30] Barnes, *Scientific Knowledge and Sociological Theory*. See also B. Barnes (ed.), *The Sociology of Science: Readings* (Penguin, London, 1974).

consultant and an assistant for two or three days. We asked for a £900 fee for that which is very reasonable in consultancy terms. The legal aid authority came back and said we could have £150. That was in a case of murder. That is a nonsense.[31]

Even where the defence is able to secure funding for an expert, the level of legal aid thus severely curtails the degree to which experts investigate defence cases. One consequence of this is that defence solicitors frequently restrict the expert's role to one of commenting on the Crown's forensic evidence rather than producing any of its own:

This will really only allow me to agree with the scientist at the laboratory, as I can only assume his work and conclusions are correct. It means that I cannot give a full consideration to the prosecution evidence and the Law Society area committee are preventing the defendant from having the benefit of a total examination and verification of the evidence, they are in effect acting as an external Judge.[32]

Lack of uniformity between legal aid committees means that some areas are more generous than others. The Home Affairs Committee cited London and Cambridge as areas in which it was fairly difficult to obtain legal aid for a forensic test. Centralization of legal aid funding may go some way to meet this criticism. However, it does not remedy the fact that HOFSS examinations may have already destroyed specimens, making it impossible for independent experts to conduct their own examinations. In theory, free-lance experts can obtain access to the HOFSS laboratories, but the nature of the HOFSS relationship with the police makes this virtually impossible. In Scotland, some police laboratories are specifically instructed not to give defence experts access to their facilities. A Home Office Research Report on *The Effectiveness of the Forensic Science Service*[33] also noted that 'in practice the police alone make use of the FSS and have, to some extent, a close relationship with it'. The Criminal Bar Association argues that

[31] HAC Report (1989), vol. ii, Minutes of Evidence, 167. [32] Ibid. 177.

[33] M. Ramsay, Home Office Research Report, *The Effectiveness of the Forensic Science Service* (HMSO, London, 1986) (Ramsay Report).

the resistance of the HOFSS to handling defence work has, in fact, become institutionalized.

Even where defence experts do gain access to HOFSS facilities, their investigations must usually be carried out in the presence of an HOFSS officer and a police officer. Since the defence may well wish to keep the findings of its forensic investigations confidential (under existing criminal law the defence is under no obligation to disclose its case before trial) their presence discourages defence use of the laboratories. The defence expert can be present at prosecution tests; the results of HOFSS tests can be made available to both the defence and the prosecution; but in the absence of confidentiality, neither proposal is attractive.

Inequality of Resources: Experts Versus Lay Objectors

In forums such as public inquiries the task of challenging the big guns re-emerges in the inequality between giant corporations, State interests, and lay objectors. This can result in a powerful alliance of interests and co-ordinated intellectual and material resources set against the disaggregated interests and resources of laypersons.

Public inquiries are administrative devices with a quasi-judicial format. Parties have the right to cross-examine witnesses, each party must adduce evidence to support its cause, and those who can afford to do so employ counsel and experts. Vested interests will have already undertaken considerable research in order to make their case more persuasive. In all probability, they will also have their own in-house pool of expertise. British Nuclear Fuels (BNFS) is a good example. At the public inquiries into the development of reactors at Sizewell B and Windscale BNFS was able to secure the services of leading experts in the field as well as its own in-house experts. The State likewise had its own experts, some of whom were employed by government establishments. Local authority resources were more limited but still more extensive than those of the lay objectors, most of whom were groups of local people or interested individuals. Whereas the findings of nuclear industry scientists were generally uncritically accepted, objectors' experts were marginalized. Parker, J., called them 'eco nuts'. Costs also proved problematic. The normal daily fee for a barrister employed at the Windscale Inquiry was about £1,000 per week; supporting legal staff cost at least £10 per hour. In some more recent inquiries, barristers have charged £1,000 per day. Such fees—indeed higher fees—are commonplace within top legal circles. Top experts are also expensive. Since inquiries such as Windscale and Sizewell may go on for weeks, months, and years, costs become crucial for lay objectors. Of the Windscale Inquiry, Wynne writes, 'Groups were faced with the daunting prospect of committing themselves to legal counsel whilst gambling on being able to raise the money later. This was not all: witnesses had to be brought in from the USA and Japan. This meant a possible debt of about £15,000 and personal bankruptcy.'[34] Objectors were forced to gather funds from personal and group donation.

At the Windscale Inquiry there was no question of objectors being publicly funded through some sort of legal aid system, an issue raised again at the Sizewell B Inquiry. The Department of Energy stated that it had no intention of yielding to appeals for public funding of objectors to the Central Electricity Generating Board's (CEGB) plan to build a pressurized water reactor (PWR) at Sizewell. Despite the fact that the CEGB was ready to provide the necessary finance, the then Secretary of State for Energy, Nigel Lawson, refused requests for funding but ordered free transcripts to be made available to objectors who found them too expensive to purchase.

Cumulatively, these inequalities of expertise and resources make the task of the objectors extraordinarily difficult. They are ranged against large-scale organizations with immense financial, material, and intellectual resources, whose experts are presumed to be authoritative and impartial. BNFS, the Atomic Energy Authority, and the Central Electricity Generating Board could combine resources in a united effort,

[34] B. Wynne, *Rationality and Ritual* (British Society for the History of Science, Chalfont St Giles, 1982).

pooling full-time legal and scientific personnel to construct the case for the nuclear industry. The eventual outcome of the inquiry may be presented as flowing naturally from the facts rather than from the interplay of social, political, and economic forces.

In the aftermath of the Windscale Inquiry, one or two groups of independent experts made themselves available to local objectors, thereby balancing out some of the inequality in scientific expertise. However, it has been argued that the increasing use of experts by both sides in such debates limits the space in which the layperson may participate.[35] It encourages the view that experts alone should frame the agenda of the debate.[36] Experts are cultural as well as legal resources. They play a significant role in the making of public policy decisions, helping the tribunal to reach a decision and defining the issues to be addressed. Issues of importance to the accused, the plaintiff, and the lay objector are marginalized. Who is selected as an expert thus becomes important, not only because the opinions of some experts will carry more weight than others, but also because some kinds of experts have more authority to arrange the agenda of debate.

Taken together, the system of access, scarcity of career opportunities, and process of socialization ensure that experts not only learn the outward appearances of their role, but also learn the cognitive and affective layers that are directly appropriate to this role: 'If socialization into the institution has been effective, outright coercive measures can be applied economically and selectively. The more conduct is taken for granted, the more possible alternatives . . . will recede and the more predictable and controlled conduct will be.'[37] The fact that expert witnesses have become very much taken for granted in the legal system is a measure of how far they have learnt the classifications and routines of the legal culture. To be partial is still a requirement of the system itself. The structure of

[35] H. Hirsch, 'Public Participation and Nuclear Energy', *Gorleben International Review* (Nov. 1978).

[36] Y. Ezrahi, 'The Political Resources of American Science', *Science Studies*, 1 (1971), 121.

[37] Barnes, *Scientific Knowledge and Sociological Theory*.

advocacy has not changed. No one in the adversarial system is paid to bring out the whole truth and nothing but the truth. Experts and lawyers are hired to do a job of work on managing information in the client's best interests. The Crown expert still has a head start in the neutrality stakes, the accused is still at a disadvantage. Blatant partisans have come to be seen as the ancient folk devils of expert witness history. The modern expert witness is presented as a team man, not so much in the vanguard of science but solidly of its centre. His countenance is uncontroversial, ordinary rather than eye-catching, his whole demeanour acquiescent, and his expression unclouded by personal interest. Moreover, the major part of his work takes place out of the courtroom and away from the public eye. He is, in a sense, an invisible player in the legal process.

Does this new image of the expert really signify that there are now fewer venal experts hoodwinking the legal system and distorting the truth? Or is it that modern day experts have simply learnt better how to present their evidence to a court of law? In this chapter I have aimed to show how some experts at least have become more skilled as witnesses. This is aided by a certain homogeneity of outlook, which in turn is produced by the procedures of access and a system of socialization into the legal subculture. Accepting a place in this subculture entails relinquishing control over one's own professional product. In return for this, experts are seldom challenged about their professional status. The downside of choosing to become a professional expert witness is the transient nature of one's participation in the legal process. This means that however much an expert participates in the backstage planning of a case, he will never be more than an auxiliary player in the structure of advocacy. Being a witness means being a bit player. It limits his power to command access to information and his control over his own testimony.

8 Processing the Legal Story: Civil Litigation

Lawyers structure cases in order to persuade. As I have shown in the previous chapter, it is important that the witnesses they use to call out the evidence must be plausible. This plausibility principle guides their actions throughout the legal process. Napley cites the ability to persuade as the one attribute which the practising lawyer must possess. It is the most fundamental of all legal skills.[1] The legal steps outlined below provide the channels by which legal fragments become persuasive accounts generated by opposing teams. Decisions have to be made about what is worth investigating and reporting at all. Doubt is constructed as much as certainty. Experts have a part to play in this process. They play it willingly or unwittingly but they play it none the less. They do so not only in the witness-box but also pre-trial.

In the process of case construction, lawyers use the formal technical rules to manage the traffic in communication both in the trial and the pre-trial stages.[2] They are guided both by substantive and by procedural laws which not only determine what they must do, but also what they might. Constructing the legal case may thus be described as an exercise in competitive lying.[3] The final version is a representation of reality; it must appear to be untouched by human hand; it must seem natural. How is this accomplished? If we believe Napley, the trick lies in the careful preparation of cases. Part of this preparation is the lawyer's selection of the appropriate tools of advocacy, including the expert. As we have seen, lawyers do not choose their expert team members at random. They select team

[1] D. Napley, *The Technique of Persuasion* (Sweet & Maxwell, London, 1975).

[2] W. D. Loh, 'The Evidence and Trial Procedure', in S. M. Kassin and L. S. Wrightsman (eds.), *The Psychology of Evidence and Trial Procedure* (Sage, Beverley Hills, Calif., 1985), 18.

[3] B. Manning, 'If Lawyers Were Angels', *American Bar Association Journal*, 60 (1974), 82.

members who can, in Goffman's terms, be trusted with the business of maintaining a given version of reality, but who will not perform their part in a self-conscious fashion.

The Legal Steps: Getting Started

Working alongside solicitors in their offices provides some insights into how cases are routinely constructed. In one sense, there is no distinction to be made between preparation and presentation: the entire process from beginning to end is one of impression management. Right from the start, what the lawyer does and how he does it will determine case outcome. The whole account has to be carefully managed from the outset. Whether the end result will be a court case, settlement out of court, or the dropping of the entire action, the initial steps remain broadly the same. These consist of the following textbook stages:

1. The File on the Desk.
2. The Letter Before Action.
3. Drafting of Statement of Claim (or Defence).
4. Request for Further and Better Particulars (or Reply to it).
5. Drafting of Defence/Amended Statement of Claim.
6. Investigation of Claim/Defence.
7. Instructions to Expert.
8. Expert's Report.
9. Exchange of Expert Reports and Case Reappraisal.
10. Case Proceeds on Course for Trial or it may Go to Sleep.
11. Negotiation of Out of Court Settlement or Case set down for Trial.
12. Settlement Out of Court or Trial.

Even though 90 per cent of civil cases settle out of court, litigation proceeds on the assumption that this assembly of papers will end up in court. The process of case construction begins many months, if not years, before cases go to trial. Generally, cases which have to go to trial in order to prove their strength are seen as failures. Some cases simply go to sleep for so many years that one is never sure whether they may still be called contested cases at all. A civil case can easily take several years to come to trial.

Case Construction: Opening Moves

Some lawyers will give the expert all the information in the legal file and ask him to come up with a version of events culled from it but few are prepared to take this risk. The second approach is to give the expert limited space in which to do damage. The space he has will be determined by the amount of information a lawyer gives him, and the kind of instructions he receives. Narrowing down an inexperienced expert's chances of damaging a legal case means giving him the minimum of information and highly specific instructions. When it comes to management and control of information, the inexperienced expert has less idea of what information the lawyer withholds. As a result, he is more vulnerable to the kinds of manipulative tactics used with less success on more experienced court regulars.

In a civil case, the first legal step after casting the legal gaze over the file is to send what is known as the Letter before Action. In his textbook on advocacy, Napley says this might better be described as a 'letter to avoid action', since this is its objective. At the same time, the lawyer must be sure that if action does follow, the letter will not hinder the case. Lawyers are advised that it should be skilfully drawn up, for it is the first legal act whereby information is sought, arguments and facts tested, claims and refutations hinted at. How well it is drafted depends on the lawyer's skill and whether, at this early stage, he has already called in an expert to advise on the contestability of the action. In the Letter of Reply, the solicitor hopes to obtain information; with luck, more information will be let slip than is asked for. Lawyers drafting the Reply seek to avoid giving away more information than they have to. 'When in doubt leave it out' is the maxim which guides all future action. It is something the expert is required to learn.

Having received a Letter of Reply, the next step is to obtain what are called Further and Better Particulars. Lawyers for the defendant seek to tie down the claimant to precise issues, whilst lawyers for the plaintiff seek to open up as many avenues as possible. For example, where a woman had slipped on some restaurant stairs and suffered serious injuries, the details requested included questions such as: was it raining the day she slipped? Which step did she slip on? Was there a handrail there at the time? What kind of shoes was she wearing? What kind of soles did they have? Given the length of time it takes a case to reach this stage, it may be impossible to remember all these details, yet failure to do so may impair the credibility of the claim. A legal case is manufactured from such fragmented remains of past events. Not all fragments are gathered. Of those which are, not all may be used. The fragments collected include correspondence with clients, with witnesses, with the other side, internal memoranda on the progress of the case, expert reports, lawyer's remarks, and counsel's advice. They may also include plans and diagrams of machinery, places, and so on. These fragments of the past are, as Garfinkel might say, tokens that will permit an assembly of an indefinitely large number of mosaics.[4] Lawyers piece them together to construct one particular mosaic, their case. They reconstruct historical fact and make the fragments speak for themselves in what seems to be a relatively unambiguous fashion. This lack of ambiguity is a social accomplishment: historical facts about events are routinely used as unambiguous resources. The scientific facts adduced to support a case are also accomplished. Latour and Woolgar have argued that modern laboratory investigation 'offers no direct window on nature. Laboratory facts are factitious, fabricated by complex

[4] H. Garfinkel, 'Good Organizational Reasons for "Bad" Clinical Records' and H. Garfinkel, 'On the Origins of the Term Ethnomethodology', both in R. Turner (ed.), *Ethnomethodology* (Penguin, London, 1974). See also D. H. Zimmerman, 'Fact as Practical Accomplishment', in the same volume; H. Garfinkel, 'The Rational Properties of Scientific and Commonsense Activities', in A. Giddens (ed.), *Positivism and Sociology* (Heinemann, London, 1974); H. Garfinkel, M. Lynch, and E. Livingston, 'The Work of Discovering Science Construed with Materials from Optically Discovered Pulsars', *Philosophy of the Social Sciences*, 11 (1981), 131–58.

programmed searches: seek and ye shall find.'[5] This applies equally to the investigations and experiments carried out by free-lance experts in legal cases. Their investigative gaze is pre-structured by legal culture and the imperatives of litigation. They may be guided by explicit or unspoken expectations and rules. But what is seen depends on who is looking and for what.

Use of an expert in drafting claims, Replies, and Further and Better Particulars ensures that statements made at the earlier stages have an air of technical competency and authenticity about them. Experts are sometimes useful before the Further and Better Particulars stage because they can anticipate what kinds of issues might be raised and can thus alert the lawyer to the need for particular sorts of questions and answers. The expert's skill lies in drafting questions; he is co-opted as interrogator, a device by which the truth of past events might be better divined. Witness skills still play an important part in this initial choice of expert, since lawyers are concerned to know that an expert will make a good witness should the case go to trial:

> We use an expert who is going to make a good witness in court and who is going to give a good, reliable report.
>
> If the expert gives the show away in the witness-box, it only increases our chances of losing somewhat!

The decision about which expert to use, and how many experts to use, almost always lies with the lawyer, although on some occasions ratification by the client is sought, especially if the clients are corporate clients with their own in-house legal and technical expertise. Where a client is legally aided, ratification for use of an expert has to be sought from the Legal Aid Fund. It will probably accept the lawyer's choice since it is the lawyer who controls the case; in Rosenthal's

[5] B. Latour and S. Woolgar, *Laboratory Life* (Sage, Beverley Hills, Calif., 1979); S. Woolgar, 'Irony in the Social Study of Science', in K. Knorr-Cetina and M. Mulkay, *Science Observed* (Sage, London, 1983); H. M. Collins, 'An Empiricist Relativist Programme in the Sociology of Scientific Knowledge', in Knorr-Cetina and Mulkay (eds.), *Science Observed.*

terms, it is the lawyer who is in charge.[6] Repeat players such as large corporations exercise a much greater degree of control over case construction. Prior processing of personal injury claims through trade unions and insurance companies generally means that these organizations have some say in which expert to appoint. They pay the bill at the end of the day and appearances require at least some form of consultation. Thus one insurer wrote to lawyers in the following terms: 'We agree to contest the case and would like you to press for a statement of claim. We suggest an expert, Mr H., to carry out an inspection. We appreciate the sum involved in this case is not huge but we have always found that expert evidence in a case of this kind is sometimes the difference between winning and losing.'

Instructions to the Expert: Sifting Information

Limits on the range of investigation are set initially by legal and financial parameters. The close relationship between lawyers and their experts means that they are frequently able to make their first contact by telephone. Much of their conversation is therefore never transcribed except by way of a memo in the file. It becomes clear, however, that good experts do not always need to be told what to do. They initiate legal possibilities. They collect passing remarks, rumours, or stories which may turn out to have some bearing on the case. Lawyers provide the editorial guidelines for case construction. What is left out is just as important as what is kept in. This equally applies to the instructions lawyers give experts. They do not always tell experts everything they need to know, nor do experts always wish it otherwise. HOFSS experts make the same complaint—the police either give them too much or too little to go on. Some experts complain that lawyers send them the whole file when what they need is a list of definite questions requiring specific answers: 'I'd rather they asked pointed questions instead of sending me all the file. Sometimes I get huge files,

[6] D. E. Rosenthal, *Lawyer and Client: Who's in Charge?* (Sage, New York, 1974).

sometimes nothing—usually from country solicitors who have no idea. The majority of my work comes from people who are used to it and send you what they think you require. Some firms don't send it all so that your mind won't be coloured before you go to court.'

An expert's investigations and reports may therefore be unwittingly based on an incomplete picture of the case. Some experts may in fact be made aware of the entire case and co-opted as a party to its ultimate construction, talking informally about the positive aspects of their case and its limitations. Likewise, scientists talking informally to colleagues may present quite a different account, in quite a different style, from the one presented to a rival, to an outsider, to an international academic gathering, or to a prestigious scientific journal.[7] Moreover, how each actor in each of these different situations interprets what he hears will, in good part, be shaped by the context in which he hears it. Gilbert and Mulkay put it thus:

> The degrees of variability in scientists' accounts of ostensibly the same actions and beliefs is, in fact, quite remarkable. Not only do different scientists' accounts differ; not only does each scientist's account vary between letters, lab notes, interviews, conference proceedings, research papers, etc., but scientists furnish quite different versions of events within a single recorded interview transcript or a single session of a taped conference discussion.[8]

The task for the lawyer and his expert adviser is to co-operate in their extraction from this diversity of accounts of the definitive version. Gilbert and Mulkay point out that scientists themselves may do this in their own work by eliminating certain statements, hyperbole, irony, rhetoric, and so on, by restating what they really meant, or 'by interpreting the data in accordance with tacit understandings gleaned in the course of interaction with participants'.[9] In the adversarial context, the tacit understandings are gleaned from past and present interactions with lawyers and the legal process. This culminates

[7] G. N. Gilbert and M. Mulkay, *Opening Pandora's Box: Sociological Analysis of Scientists' Discourse* (Cambridge University Press, Cambridge, 1985).

[8] Ibid. 11.

[9] Ibid.

in a knowledge of what kinds of interpretative uncertainties must be left out.

The Disclosure and Control of Information

Lawyers thus employ a variety of strategies to structure expert evidence pre-trial. They may control the flow of information an expert receives, they may provide a restricted schedule of issues to which the expert is directed to apply his mind, and they may employ a variety of strategies to keep the expert in the dark regarding the full legal story. Thus, whilst the expert himself may think that he comes into the legal process with an open mind, he is often unaware of the degree to which the information he is given has already been shaped by adversarial strategies. Whether he likes it or not, this will affect what he looks for, how he looks for it, what he finds, and what, ultimately, lawyers make of what he finds. The expert has no power to command full sight of the file; he is not asked to pursue all avenues of inquiry. He is not asked, for example, to investigate lines of inquiry which might confirm another outcome.

According to books on law, procedure, and Practice Directions, there are a number of legal rules which, on the face of it, appear to favour norms of openness rather than norms of secrecy in the litigation process. Thus, for example, the Rules of the Supreme Court (RSC) state that 'Where in any case or matter an application is made under Order 36(1) in respect of oral expert evidence, then, unless the Court considers that there are special reasons for not doing so, it shall direct that the substance of the evidence be disclosed in the form of a written report or reports to such other parties and within such period as the Court may specify.' The effect of Order 37 is, according to one commentator, 'to create a strong presumption of pre-trial disclosure where the parties seek to call expert evidence at trial'.[10] It has been argued that this rule, along with amendments to the Rules of the Supreme

[10] T. Hodgkinson, *Expert Evidence* (Sweet & Maxwell, London, 1990), 135.

Court in 1987,[11] have made the legal process more open by restricting lawyers' opportunities to hoodwink each other in the pre-trial process of case construction. Hodgkinson argues, for example, that 'the very broad measure of discretion as to disclosure of expert evidence generally has been narrowed considerably so that there now must be "special reasons" for not ordering disclosure'.[12] Moreover, personal injury cases have been put under a separate automatic directions procedure[13] and the rules which formerly made medical negligence cases subject to a special position as regards disclosure have now disappeared.[14] Furthermore, case-law now suggests that, where disclosure has been ordered, the parties to an action must disclose the substance of the report, and this has been taken to mean the entire report compiled by the expert.

However, judicial disapproval may appear to control sharp practice, but it has a very limited impact. The law in the books permits lawyers to engage in a number of perfectly legitimate strategies designed to keep secret information which, in an adversarial setting, is potentially damaging to one's case. One question which arises therefore is whether, and how far, this strong presumption of pre-trial disclosure guides normal legal practice, bearing in mind that, since only some 10 per cent of civil cases ever do proceed to trial, for most of the time oral expert evidence is not part of the lawyer's agenda. Moreover, although RSC Order 38, Rule 36, requires an application to be made to the court for appropriate directions as a condition of calling expert evidence at trial, it also provides that 'parties may agree to adduce such evidence at trial without seeking directions'.[15] Whilst this will usually result from an agreement by both parties to disclose voluntarily to each other the reports of their experts,[16] such a voluntary arrangement is by no means inevitable. Further, the rules do not affect the existing law of privilege.[17] If an expert's report is

[11] Ibid. 38. [12] Ibid. [13] Ibid. [14] Ibid.

[15] RSC o. 38 r.36; Hodgkinson, *Expert Evidence*, 49.

[16] Hodgkinson, *Expert Evidence*, 49.

[17] See *Worrall* v. *Reich* (1955) 1 QB 296; *Causton* v. *Mann Egerton (Johnsons) Ltd.* (CA) (1974) 1 WLR 162.

prepared for the purpose of existing or anticipated litigation it is a privileged document and need not be disclosed.

According to case-law, if the report is disclosed it must include the substance of the expert's report, that is, both the factual and the opinion sections. But according to Phipson, *On Evidence*,[18] the word 'substance' has itself become ambiguous. Lawyers attempt to circumvent full disclosure by disclosing only the factual part of the expert's report, rather than the full report which also includes expert opinion on the facts in issue. Only disclosing the factual part of the expert's report means that a conscious effort has been made to leave out any conclusions as to the defects of machinery, the system of work, or other relevant opinion evidence. As one judge has said, this seems to be

> a total misconception of the ordinary meaning of the word 'substance'. It is also a misconception of the function of the expert. An expert, unlike other witnesses, is allowed because of his special qualifications and/or experience to give *opinion* evidence. It is for his opinion evidence that he is called, not for a factual description of the machine or the circumstances of the accident, although that is often necessary in order to explain and/or justify his conclusions. When the substance of the expert's report is to be provided, that means precisely what it says, both the substance of the factual description of the machine and/or the circumstances of the accident and his expert opinion in relation to that accident, which is the justification for calling him.[19]

This opinion as to what constitutes the substance of the expert's report seems to impose a higher duty on the parties to disclose every single piece of information they have, including anything which is detrimental to their case.[20] However, this concern with greater candour in disclosing the real meat of the expert's report (the opinion) masks the fact that, at one and the same time, in law there is still no general obligation on the parties to an action to disclose all their factual case.

[18] S. L. Phipson, *On Evidence* (13th edn. Sweet & Maxwell, London, 1982); Hodgkinson, *Expert Evidence*, 110.

[19] *Ollet* v. *Bristol Aerojet Ltd.* (Practice Note) (1979) 1 WLR 1197, per Ackner, J.

[20] Hodgkinson, *Expert Evidence*, 110.

Meetings without Prejudice: Jointly Agreed Examinations, Inspections, and Reports

There are a number of points during the legal process where controls on information are especially important. Damaging information might be leaked by the expert, for example, in the mutual meeting of experts during medical examinations of injured claimants or visits to the scene of an incident for site inspections. These meetings are usually carried out after the expert has seen the legal documents and knows what to look for. Joint inspections or examinations with the expert from the other side are relatively infrequent, though there is a growing tendency on the part of the courts to make them a compulsory part of the pre-trial procedure.[21] Amendments to the Rules of the Supreme Court during the 1980s along these lines were designed ostensibly to develop greater openness in pre-trial procedure,[22] and were described as a significant step towards this end.[23] In fact, the main benefit of joint meetings is the saving of time and money. It gives experts an opportunity to iron out their disagreements, agree what they agree about, agree what they disagree about, and prepare a joint statement for trial outlining their points of contention. The procedure reflects one interpretation of the expert's duty as being to limit in every way possible the contentious matters of fact to be dealt with at trial.[24] Lawyers themselves are highly ambivalent about pre-trial get-togethers amongst expert witnesses. Experts may leak information to the other side; they may try to narrow the gaps between them where it is sometimes more beneficial to the client's case to keep these gaps as wide open as possible. Despite changes to procedure in favour of more joint discussions between experts before trial, lawyers therefore remain wary of this process. In a sense, the law's greater emphasis upon pre-trial meetings between experts makes it even more

[21] RSC o. 38 r.(3) allows the court to direct a meeting between experts 'without prejudice'. See also Hodgkinson, *Expert Evidence*, 51.

[22] See RSC (Amendment no. 2) (1986) (SI 1986 no. 1187); see also RSC (Amendment) (1987) (SI 1987 no. 1423).

[23] *Supreme Court Practice* (1991), i. 38-7, 39-8.

[24] *Graigola Merthyr* v. *Swansea Corporation* (1928) 1 Ch. 31, per Tomlin, J.

imperative for lawyers to hire the right expert at the start. An expert may be chosen for his ability to tune his eyes and ears to legal relevancies, but he is also selected for his ability to keep his mouth shut.

Specific Strategies

1. The two-part report

Quite apart from recruiting better expert witnesses, lawyers have also employed several other strategies to enable them to circumvent the spirit of greater openness. The first has been to pay more attention to the way in which the expert's report itself is composed. This involves making sure that its style and format make it suitable for disclosure. As I have already pointed out, lawyers may ask the expert to submit a report which is split into two parts, the factual evidence and the expert's opinion based on these facts. For example, some experts end their reports with the words 'We hope the report is in an exchangeable form.' Some mark the factual section of their report 'for disclosure', indicating that they are fully aware of the use to which their reports are put and have integrated this into their style of presentation. Despite judicial disapproval this practice persists. Lawyers may, for example, exchange the first part of the report and wait to see if the other side spots the omission and asks for the remainder. If they fail to do so, some adversarial advantage may well result.

2. Sequential exchange of reports

Moreover, though in principle the exchange of reports is generally supposed to be simultaneous, in practice it may often be sequential. It is thus possible to delay the exchange of one's own report until one has obtained the other side's report. This may enable lawyers to go back to their own expert and, as one lawyer put it, ask him to doctor his report in the light of information received. Errors and omissions can thus be corrected without anyone but those inside the legal team ever becoming aware of them.

3. No duty to disclose all reports

Furthermore, the law by no means requires that all expert reports be exchanged. Lawyers are required to exchange only those reports upon which they intend to rely at trial. For every one report disclosed in this way there may very well be others which remain undisclosed, usually because they do not support the client's case. The exact extent and range of expert disagreement, and the degree to which opinions in the report relied upon concur with (or deviate from) those of the profession more generally, is thus never revealed to the court.

4. Agreeing to disagree

The need to agree expert reports before trial at one point led lawyers to 'agree to disagree'. In effect, where experts could not agree, the lawyers would comply with procedural requirements by simply agreeing what were, in fact, quite conflicting reports. The point of trying to agree expert reports pre-trial was to cut down the contested issues and reduce time and costs. 'Agreeing to disagree' thwarted these aims. It resulted in cases going to trial in which the issues remained unresolved. Trial courts were faced with the perplexing problem of having to conduct lengthy trials where what had been anticipated was a much shorter contest. The fact that lawyers were only nominally complying with this procedural rule led the judiciary to seek an administrative order putting a stop to this practice.

5. The covering letter

In addition to the two-part report, experts typically send a covering letter to the lawyer which includes information not in the report itself. The covering letter has a long history in legal practice. It found renewed favour under these conditions of expanded procedural openness. Essentially, it is a device for circumventing disclosure of damaging information. The covering letter is aptly named. It not only contains any information which might, if included in the report, damage the case under construction, it also helps the expert and the lawyer to cover up any trace of disconfirming evidence. Experts are encouraged to present a covering letter partly because, unlike the report, the covering letter is not subject to disclosure. It is standard practice to leave damaging information out of the report because it might be exchanged with the other side. According to the rules of advocacy, it is the experts' job to alert their own side to such information; it is not their job to make out a case for the other side.

The practice of using covering letters as a means of avoiding the disclosure of damaging information drew judicial comment in the case of *Kenning* v. *Eve Construction* in 1989.[25] In this case, the expert sent a covering letter along with his report to the solicitor. By mistake, the solicitors disclosed the letter to the plaintiff's solicitors. The court noted that the practice of using such letters was widespread and decided that they were not immune from disclosure.[26] This decision was a complete departure from previous practice and caused some consternation. It seemed to remove the protection of professional privilege from the expert's covering letter and place a duty on the expert to reveal to the court matters which were contrary to his client's case.[27] However, whilst judges may occasionally imply that experts have a higher duty to the court than to the client employing them, there is a conspicuous absence of any legally binding rule to this effect, nor are experts themselves liable to civil or criminal action should they choose not to reveal all they know to the court. Moreover, a year after *Kenning* v. *Eve Construction*, in 1990, the Court of Appeal in *Derby* v. *Weldon (No. 9)* ruled that Order 38, Rule 37, of the Supreme Court Rules required that the expert witness should disclose only the evidence that it was intended he should give at trial.[28] Staughton, LJ, ruled that in so far as *Kenning* v. *Eve Construction* was an authority to the contrary, it had been wrongly decided. The covering letter and its protection under legal professional privilege thus remain intact.

[25] *Kenning* v. *Eve Construction* (1989) 1 WLR 1189.
[26] Hodgkinson, *Expert Evidence*, 91–2. [27] Ibid.
[28] *Derby* v. *Weldon (No. 9)* The Times, 9 Nov. 1990.

6. *Constructing the Expert's Report*

The expansion of practice rules requiring greater openness seems to have proved counter-productive. Far from becoming more open, lawyers have found new, behind-the-scenes, ways of avoiding disclosure which are more intense but less visible than before. One symptom of this trend is the increasing influence of lawyers over the form and content of the expert's report. There are no general legal rules as to the form of the expert's written report,[29] though Hodgkinson advises that the expert should draft the report 'in a manner appropriate to its forensic purpose'.[30] He goes on to write of the expert's duty that it is:

> not to represent the interests of his client, but to express his views honestly and as fully as is necessary for the purposes of the case. A combative tone is inappropriate and only serves to undermine the likelihood of its conclusions being accepted. Criticism of the personality or conduct of other expert witnesses is not justified unless clearly relevant and undeniably supported by the facts. The expert should attempt not to allow his opinions to be coloured by sympathy for his client or his client's case.[31]

However, Hodginkson goes on to explain in what sense an expert should allow himself to be guided by the imperatives of the legal process:

> There is a wholly distinct sense, however, in which an expert can and should assist his client's case. His report can be presented in a form in which its conclusions, which represent the expert's true opinion, are most likely to be readily accepted by the tribunal which is to hear the case, or which may cause the other parties' experts and legal advisers to reconsider their current stance. Its form may also assist his client's advocate in putting the merits of his case to the court . . . First, the report should recognise the shape and emphasis of the client's case.[32]

Hodgkinson suggests the expert start with the pleadings, a legally drafted account which 'usually sets out the case in as persuasive a manner as possible'.[33] He should also start with

[29] Hodgkinson, *Expert Evidence*, 84–5. [30] Ibid. [31] Ibid. [32] Ibid.
[33] Ibid.

the stronger points in the case, as this 'facilitates the presentation of the case at trial'. He should avoid using legal expressions (such as 'negligent' or 'breach of contract') and should be wary of using adjectives such as 'defective' or 'inadequate'. Yet quite frequently the questions posed by lawyers involve these issues of negligence and breach of duty; some touch upon the very issue which in court the judge and jury are supposed to decide. In courts of law experts may be challenged on these grounds but in the pre-trial process there is no one to challenge them. Nor is the practice illicit. Indeed, lawyers seem quite astonished at the suggestion that experts are not supposed to do this kind of legal work, arguing that 'that's what experts are for'; 'They always report on such matters'; 'always deal with legal concepts like reasonableness and balance of probability'; 'determine factual issues and assess damages and liability.' They 'just have to pick up the facts and say, from their experience, what are most likely to be the facts of the matter in this case'.

The expert's report of his investigations is a particular kind of document produced for specific purposes (the adversarial contest), with a peculiar lexicon of its own. Given this, and given the trend towards greater disclosure, it is hardly suprising that lawyers have increasingly sought to influence both the content and form of the report in order to increase its suitability for disclosure. It has long been counsel's practice, for example, to instruct the solicitors to settle the expert's report in cases where the report is of great importance and not suited to the needs of litigation.[34] As Hodgkinson notes, there is also little doubt that 'the practice has also been used in order to improve upon the report by excision and amendment, or the use of descriptive words which have some direct relation with the legal matters which require proof'.[35] In practice, barristers are professionally entitled to settle the expert's report.

7. *Editing*

Sometimes experts fail to omit damaging information. In these circumstances, they may simply be asked to alter their reports

[34] Hodgkinson, *Expert Evidence*, 84–5. [35] Ibid.

in order to omit certain unfavourable evidence. The following comments give a flavour of what is requested and experts' reactions to it:

Solicitors tell us what they want us to say, and we are quizzed as to what we would say and if we would accept such and such. On one occasion the solicitors went through the papers and talked to us, which amounted to vetting us to see if we would give the right sort of evidence. This was for about three hours.

Some try; I don't take any notice usually.

They tell you what they would like to hear but if they did try to tell me what to say I wouldn't take any notice.

I've heard of it happening. It's sometimes suggested that you omit certain things, though don't tell lies. A report will be exchanged but a covering letter won't so you put things against your side in this.

I've never been asked to restructure my report but I would be prepared to agree if I thought my report showed a bias or . . . created an unfair impression.

Yes, solicitors ask me to restructure. I take no notice.

If they did, I wouldn't be guided by it. There are times, however, where I may be asked to put my evidence in a different *way*. I am usually prepared to agree.

It is difficult to find examples of explicit directions to edit reports. Usually, requests to edit are made by telephone rather than in writing and are thus difficult to document. But an examination of solicitors' files does reveal some examples. In one case, an expert was asked to redraw a map of a town in order to leave out the part which showed a lorry park. The defendant in the case had parked overnight in a side street, colliding with a car. His action in parking here was felt to be less justifiable once it could be shown that he could have parked his lorry in the proper place. Given such an opening, it was felt that the other side would waste no time pointing out that the lorry driver was at fault and had not given a totally honest account. In another case, counsel wrote to the solicitor advising that: 'The last paragraph of the first page of the report contains a statement which is unfavourable to the plaintiff's case and I do not see why we should hand this to the other side. There is no need for this statement to appear in the report. Accordingly I advise that the report be recast so as to omit that paragraph.' In yet another case, involving a fire at sea, the expert was asked to omit a paragraph in which he observed that some metal parts looked as if they had been hammered off, the implication being that an exit route which should have been open in fact had to be forced open.

Despite judicial disapproval editing remains widespread. One commentator advises that members of the Bar 'must use their discretion as to the extent to which they preserve the distinction between improving the expression of an opinion and changing its meaning'.[36] If changes are made, the Bar Code of Conduct recommends that the expert be shown the altered report, but there is still no guarantee that this will happen, nor does the expert have any *locus* to insist. Moreover, whilst some lawyers are reluctant to admit to editing, others argue that they would be negligent if they did not edit out damaging information, that is, if they allowed something prejudicial to their client's case to go into the expert's report and thence to the other side. They argue that their duty to the client entails constructing the best possible case on his behalf, whilst, as we have noted, the norms of advocacy place no duty upon the lawyer to make the other side's case out for them.

Editing also indicates how dependent experts are upon lawyers for their information about what the legal process requires of them. If they are told that there is no need to include a particular piece of information in a report because it is not legally relevant or because it is hearsay, they are in no position to contradict this advice. This is particularly true for novice experts. Experienced experts, on the other hand, generally need no instructions about what to put into their reports and what to leave out. They have been successfully educated by lawyers in the norms of the adversarial process. Some textbooks advise experts to consult with the solicitor and counsel before preparing the final report, in order that some guidance might be given 'as to how the report is best presented. Every

[36] P. S. D. Reynolds and M. P. King, *The Expert Witness and his Evidence* (BSP Professional Books, Oxford, 1988), 38.

endeavour should be made to comply with counsel's wishes as to any particular format or order he requires.'[37]

8. Reporting Styles

The same text also advises experts on their reporting style: 'Most reports begin with the name, address, qualifications and experience of the person writing the report. It is likely that a number of drafts will be prepared and modified prior to the final version being produced for exchange with the opposing expert. It is good practice in early drafts to make reference to any weaknesses in the argument being put forward, together with an indication of their significance.'[38] Different experts have different models which they use in order to compile their reports. These are generally drawn from their experience of report writing within their own professions. They rely upon their scientific, technical, or medical training for a format. Some models are deemed by lawyers to be more suitable than others. The model lawyers prefer divides reports into fact and opinion sections. Factual information generally includes the name, address, and age of the plaintiff, and details of any injuries caused by the accident. Medical and social work reports sometimes include a synopsis of the plaintiff's social situation and details of any previous medical history. All reports contain some explanation for their making.

An aura of neutrality in report writing is crucial to the achievement of credibility in court. The more technical and impersonal the language and tenor of an expert's report, the more it appears to be unbiased. The best reports therefore must appear seamless, unaltered, and impartial. They must show no traces of editing or selectivity. Expert reports seldom go into detail about the methods chosen for testing or investigating a problem, and never discuss the limits of such techniques. They rarely mention disconfirming results of tests or investigations. The style adopted by good experts mimics the conventionalized impersonal style of scientific reports. It makes, 'the author's personal involvement less visible . . . and

[37] Ibid. 15. [38] Ibid.

the existence of opposing scientific perspectives . . . tends either to be ignored or depicted in such a way which emphasises their inadequacy, when measured against the purely factual character of the author's results'.[39] Gilbert and Mulkay argue that this formal style gives the report's findings an aura of objectivity. It omits reference to the dependence of experimental observation on theoretical speculation and the degree to which experimenters are committed to specific theoretical positions. The overall impression of impartial appraisal is constructed by recourse to general formulas which give the impression that the application of methodological procedures is a highly routinized activity, with little room for individual initiative and variability.[40] As Latour and Woolgar write: 'The result of the construction of a fact is that it appears unconstructed by anyone; the result of rhetorical persuasion in the agnostic field is that participants are convinced that they have not been convinced.'[41] Hunston calls this an ideological inversion. The way in which experiments are written up in normal scientific settings is in keeping with the non-personal ideology of science.[42] Any reference to the choices and judgement of the scientist is kept to a minimum. Karen Knorr-Cetina has shown how the means employed by scientists to research the physical world themselves construct the picture observed.[43] The way in which scientific inscriptions (figures, tables, graphs, laboratory notes, etc.) are transposed into results further constitutes this picture, as does the way in which these results are then written up in the scientific article or report. The cumulative effect of these practices is to convince us that the facts have spoken for themselves. This being so, it is very hard to argue against whatever they say.

Hunston has pointed out a number of linguistic strategies adopted by scientists in their written reports which aim at convincing the reader of their worth. Stylistically, scientific writing

[39] Gilbert and Mulkay, *Opening Pandora's Box*, 47.

[40] Ibid.

[41] Latour and Woolgar, *Laboratory Life*, 240.

[42] S. Hunston, 'Evaluation and Ideology in Scientific Writing', in M. Ghadessy (ed.), *Varieties of Written English*, vol. ii (Pinter, forthcoming).

[43] K. Knorr-Cetina, 'The Ethnographic Study of Scientific Work', in Knorr-Cetina and Mulkay (eds.), *Science Observed*, 115–41.

has no attitudinal language. It is apparently free from personal judgement and objective.[44] However, she points out that it would be wrong to characterize the writing as non-evaluative,[45] and 'In spite of the absence of such attitudinal lexis as good, excellent, successful, the writer's attitude to the value of the research is clear.'[46] Scientific writing thus carries implicit messages. The choice of words such as possible, probable, likely, showed, pointed out, speculated, claimed, found, appear, seem, suggest, and so on pushes the certainty of the writing up or down. The use of such techniques in scientific writing means that the expression of value is often inexplicit. This inexplicitness 'leads to the common judgement of scientific writing as impersonal and non-evaluative'.[47] The role of the scientist in interpreting the data is suppressed 'so that the results themselves are made to appear responsible for the conclusions'.[48] Authors 'construct texts in which the physical world seems regularly to speak, and sometimes to act, for itself. Empiricist discourse is organized in a manner which denies its character as an interpretative product and which denies that its author's actions are relevant to its content.'[49] Science (and law) is dominated by what Mulkay and Gilbert have termed an empiricist repertoire which displays the kind of recurrent stylistic, grammatical, and lexical features noted by Hunston.[50] The production of a scientific text is a social process. It represents an interaction between the writer and the reader; it is a text which plays a role in a particular social system.[51] Because social systems incorporate ideologies, the text 'is therefore written to be understood within the context of a particular ideology'.[52] On this view, scientific writing is not simply a report but an attempt to persuade the scientific community to place a high value on the scientist's knowledge claims.[53] As Hunston puts it:

[44] Hunston, 'Evaluation and Ideology'. [45] Ibid. [46] Ibid. [47] Ibid.

[48] M. A. K. Halliday, *An Introduction to Functional Grammar* (Arnold, London, 1985). See also M. A. K. Halliday and R. Hassan, *Language in a Social-Semiotic Perspective* (Deakin University Press, Victoria, 1985).

[49] Gilbert and Mulkay, *Opening Pandora's Box*, 55–6. [50] Ibid.

[51] Hunston, 'Evaluation and Ideology'. [52] Ibid.

[53] G. N. Gilbert, 'The Transformation of Research Findings into Scientific Knowledge', *Social Studies of Science*, 6 (1976), 281. See also B. Latour, *Science in Action:*

No information presented in a research article is neutral with respect to the value system. Rather, the entire article rests on, and is interpreted in the light of, an evaluative sub-text of assumptions and comparisons . . . The value-system of the target community must be absorbed and information and argument must be presented in its terms. The final product must be expressed in a way that both says what the [scientist] wants to say and fits what the reader expects to hear. The ideology of the discipline must be conformed to, yet its value system must remain implicit.[54]

The apparently pre-ordained objectivity of scientific results and reports is thus shown to be itself both a construction and constituent of the cultural authority of science. Science and law may comprise different systems of authority but they use identical means to achieve their goals. Science, like law, silences the interpretative and contingent nature of its methodology and its results.

However, scientists interviewed by Gilbert and Mulkay informally stressed the craft, skills, subtle judgements, intuitive understandings, and the inventiveness of their approach to investigation. These are exactly the kind of skills which lawyers look for in a good expert witness. Equally if not more important is the writer's ability to erase or silence all trace of these qualities so that his report may become a highly abstract version of his activities.[55] The value of this style of reporting for the lawyer is that it disguises the contingent nature of the case and makes the facts seem self-evident and non-negotiable. Categories of presentation reflect the clear-cut, unproblematic nature of facts which speak for themselves. Such techniques do not simply reflect reality: they shape it. Allied to this are formal ways of presenting evidence which induce conformity and diminish not the human factor itself but its visibility. The eventuality of the published report pervades the organization of inquiries all along their course.[56] A standardized version is

How to Follow Scientists and Engineers Through Society (Open University Press, Milton Keynes, 1987); H. M. Collins, *Changing Order: Replication and Induction in Scientific Practice* (Sage, London, 1985).

[54] Hunston, 'Evaluation and Ideology'.

[55] Gilbert and Mulkay, *Opening Pandora's Box*, 55–6.

[56] Latour and Woolgar, *Laboratory Life*. See also T. J. Pinch, 'The Sunset: The

produced which can withstand scrutiny within and without the profession. The customary form of reports allows the expert to say what he must but not what he might.

The Impact of the Law

The process of disclosure is supposed to be one in which each side reveals its hand. It is supposed to be the one device of due process which allows the accused party to discover the case against him. In practice, disclosure and discovery are two different parts of a cat-and-mouse game. The discovery process affords the lawyer an opportunity to silence or manipulate information. For example, the writer of one memo says, 'We note from your letter that you do not admit to having a report by Dr X.' Internal memos in lawyers' files reveal that parts of reports have been deleted. In the absence of positive rules outlawing this kind of backstage trick it is quite likely that only some of the evidence will be disclosed because only some of it is discovered. Disclosure is therefore never a very open and above board procedure. It is another stage at which the expert may be required to play a man of law role. One text advises the expert that:

> He must know his own side's strengths and weaknesses and decide what he wants to obtain, or should be looking for, in the other side's documentation. He will want to discover and investigate the documentation not only for trial purposes but also for the purposes of negotiation. He must be careful to examine the other side's documentation to ascertain what evidence supports his client's case and what evidence is against it.[57]

The expert's tasks in this stage of proceedings include 'using the evidence to the best possible advantage of the client' and generally assisting the legal team.[58] He is advised to ask himself whether there was other evidence which he has not obtained but which the other side should or might have and what, out of all the evidence, should be referred to in the final

Presentation of Certainty in Scientific Life', in H. M. Collins (ed.), *Social Studies of Science*, 11 (1981), 131–58.

[57] Reynolds and King, *The Expert Witness*, 93–4.

[58] Ibid. 102–3.

report. He should also ask himself whether a document can stand alone to prove a fact or whether it requires corroborative evidence, whether it is useful to the client's case, and whether it is admissible. Given this man of law role it is not surprising that direct recasting of expert reports is sometimes seen as imperative if they are to be exchanged. In any event, lawyers are under no legal obligation to exchange all the reports they have obtained. As one lawyer put it, 'There's no point in concealing advantageous reports but there is a great deal of point in withholding disadvantageous ones.' Reynolds and King advise experts to ensure that their preliminary reports are regarded as privileged documents, for should they be revealed, the expert might be embarrassed by proof that he has changed his mind.[59] Similarly, experts are advised to consult with solicitors about the possibility of restricting circulation of the report within the office and making sure that memoranda and notes are covered by legal professional privilege. Though the status of the covering letter as a privileged document has been challenged in the courts, its content appears to remain safe from disclosure.

One case which illustrates how disclosure and discovery may be manipulated involved a claim by a market gardener against the local Lord of the Manor. The action was for polluting river water drawn from the river downstream of the manor house to irrigate crops in the market garden. Oil from the manor house central-heating store was thought to have seeped into an underground stream, and thence into the river. A year's crop had been destroyed. The market gardener sought compensation for his loss and an expert report on crop failure was produced to support his claim that the damage had been done by central-heating oil. The defendant's solicitors sought an expert report on the crop damage, another expert report on central-heating oil storage, and two expert reports upon rainfall and river flow during the period in question. In due course all these reports were completed. Their conclusions tended to support the plaintiff's claim. This did not, however,

[59] Reynolds and King, *The Expert Witness*, 104–5.

lead to the claim being settled immediately in the plaintiff's favour. The defendant solicitors informed the plaintiff solicitors that a number of expert reports had been commissioned; they did not mention their contents. The plaintiff withdrew his claim because he could not afford to match the number of expert reports; he assumed that the defendant would not have spent so much money on expert evidence unless it supported his case. The actual content of the reports themselves was never disclosed. The case was won simply by force of numbers even though, as the defendant's solicitors commented, 'One doesn't like to win cases by such means.' However, as long as what the lawyer does falls within the letter of the law, it is permissible to win cases in what are essentially underhand ways; indeed, one's duty to the client may make it imperative.

In 1981 the House of Lords decided, in the case of *Whitehouse* v. *Jordan*, to comment upon the practice of lawyers helping experts prepare their reports: 'While some degree of consultation between experts and legal advisers is entirely proper, it is necessary that expert evidence presented to the court should be and should be seen to be, the independent product of the expert, uninfluenced as to form or content by the exigencies of litigation.'[60] Some commentators have taken this to mean that the lawyer's influence over the preparation of reports has been curtailed.[61] However, the net effect has not been to stop lawyers attempting to influence expert reports but to make them more circumspect about the means they use to do so. If the import of *Whitehouse* v. *Jordan* is that the format of the expert's report should be left to his own discretion, one implication is that lawyers will be more careful than ever to choose experts upon whose discretion they can rely. Given the anxieties of novice experts about the prospect of giving evidence it is also somewhat disingenuous to imagine that they will not follow the lawyer's advice. They will tend to copy existing report-writing conventions and, unless they are aware

[60] *Whitehouse* v. *Jordan* (1981) 1 All ER 67. See also *Kelly* v. *London Transport Executive* (1982) 2 All ER 842, 851, per Lord Denning, MR.

[61] M. S. Leigh, 'The Legal Viewpoint', in M. A. Foy and P. S. Fagg, *Medico-Legal Reporting in Orthopaedic Trauma* (Longman, London, 1990).

of the ruling in *Whitehouse* v. *Jordan* (which is highly unlikely), they will never suspect that they have any *locus* at all to resist the lawyer's advice.

The case of *Noble* v. *Robert Thompson* (1979) is an exception. It was brought by the Medical Protection Society, as a test case, on behalf of a consultant psychiatrist.[62] The issue was whether a divorced mother who had been treated for manic depression should have access to her children restored to her by the court. The contentious passage in the expert's report read, 'Access should not be enforced against the wishes of the children themselves, who are in a position to express their views.' The solicitors regarded this as prejudicial to their client's case and asked the psychiatrist to delete it. The psychiatrist refused, whereupon the solicitors refused to pay his fee, arguing that he was in breach of contract. As the *British Medical Journal* wrote at the time, the case raised the issue of whether an expert's report was supposed to be an impartial account for the court, or whether it was meant to buttress one party's case. The case resulted in a finding for the psychiatrist but, as the *BMJ* pointed out:

> One certain consequence of this judgment will be that in future solicitors who want to avoid having reports that they commission spiced with unwelcome or prejudicial medical observations will tend to give doctors more precise instructions. The commentator in the *Law Society Gazette* has already made this point. Doctors who do not want to become the unwitting mouthpieces of partisan interest will need to consider carefully the broader ramifications of each case before agreeing to present a limited view.[63]

Hodgkinson suggests that where solicitors do not wish the expert to include such items in his report 'this should probably be achieved by a full and precise letter of instruction to the expert concerned rather than by excision after the event'.[64] *Noble* v. *Robert Thompson* is interesting because it is unusual—experts do *not* normally challenge legal practices, they do not

[62] *Noble* v. *Robert Thompson* (1979) Law Society Gazette 1060.
[63] *British Medical Journal*, 24 Nov. 1979.
[64] Hodgkinson, *Expert Evidence*, 348.

see it as their business to challenge the rules of the game. The exception, then, proves the rule.

Noble v. *Robert Thompson* also focused attention on the issue of how much experts are expected to disclose. According to Sir Roger Ormrod, experts should frankly disclose to the court all the information of which they are possessed, including information which may be detrimental to their own side.[65] However, there is no clear duty on an expert in this regard, although in *R.* v. *Maguire* (1992) it was held that the failure of the prosecution's scientists to disclose material before or during trial was a procedural irregularity and 'that where the material ought to have been disclosed the irregularity would usually be material'.[66] In *R.* v. *Maguire* the prosecution's experts had failed to disclose material to the prosecution lawyers. The Court of Appeal decided that there was:

> no reason to distinguish between members of the prosecuting authority and forensic scientists providing advice to that authority, and a forensic scientist who was an adviser to the prosecuting authority was under a duty to disclose to that prosecuting authority material of which he was aware and which might have some bearing on the offence charged and the surrounding circumstances of the case; accordingly, a failure by a prosecution expert witness to disclose material was capable of constituting a material irregularity.[67]

At first sight this decision may seem to clarify the expert's role: he is to disclose material which he is aware might have a bearing on the case. However, this raises further problems. For example, how is a man of science to know what kinds of material might have a bearing on the legal case? What if the expert is unaware that information in his possession is of significance to the other side? Moreover, this decision clearly states that, for the purposes of the prosecution's duty to disclose, there is no reason 'to distinguish between members of the prosecuting authority and those advising it in the capacity of a forensic scientist'.[68] The scientist is to be regarded as part of the

[65] Ormrod, 'Evidence and Proof'.

[66] *R.* v. *Maguire* (CA) (1992) 1 WLR 767–84; *Anne Rita Maguire and Others* v. *R.* (CA) (1992) 94 Cr. App. R. 133; *R.* v. *Maguire* (1992) 2 All ER 433.

[67] Ibid. [68] Ibid.

prosecution machinery, bound by the same duty to disclose. Thus a decision which at first sight appears to promise greater independence for the prosecution expert actually defines him quite clearly as a man of law. The Court of Appeal in *Maguire* also made it clear that the duty to disclose is only a duty to disclose to the authority retaining the expert. That authority must, in turn and subject to certain provisos, disclose it to the defence. The decision does not require the expert to volunteer such material to the defence or to the court on his own initiative. Neither does the law require experts to tell the court about the methods chosen (and not chosen) in scientific tests, their limits and drawbacks, and so forth. Any duty of candour[69] thus lacks substantial legal authority. The issue of disclosure in civil procedure was raised in four cases in 1987.[70] In its consideration of these cases, the Court of Appeal argued that whilst the English courts adhered in the main to adversarial procedure they had moved away from allowing purely tactical considerations to inhibit disclosure. Using a gaming analogy, the Court of Appeal argued, 'Nowadays the general rule was that while a party was entitled to privacy in seeking out the cards for his hand, once he had put his hand together the litigation was to be conducted with all cards face up on the table.'[71] The court decided that the substance of the expert reports in these cases should be disclosed pre-trial. Although it argued that this would lead to true justice, its version of true justice seems to have been highly coloured by the need to save time, cut costs, and have the dispute settled out of court rather than by judicial decision.

Even where there is full disclosure there can be no guarantee that this leads to substantively fairer outcomes. Rules which encourage more open discussion in fact seem to have produced an increase in low-visibility behind-the-scenes manipulation of information which makes selections less discernible. This is

[69] See the comments of Lord Donaldson, MR, in *Naylor* v. *Preston Area Health Authority* (1987) 1 WLR 958, 967.

[70] *Naylor* v. *Preston Area Health Authority* (1987); *Thomas* v. *NW Surrey Health Authority* (1987); *Foster* v. *Merton and Sutton Health Authority* (1987); *Ikumelo and Another* v. *Newham Health Authority* (1987) The Times, 14 Apr. 1987.

[71] Ibid.

particularly so in the case of editing. The fact that there is rarely any need for lawyers directly to instruct experts to alter, structure, or otherwise tamper with their reports does not mean such structuring does not take place. For experts with long experience of working with the law, these tactics may have become so routine, so much part of the culture within which they work, that they become quite unconscious. Moreover, expert witnesses will rarely, if ever, see the whole legal story. Worryingly, very few seem to realize it.

The rules promulgated in the 1970s and 1980s may have been designed to promote greater openness, but clearly, one effect has been to make lawyers pursue ever more innovative and evasive tactics. This is not simply a case of the law in practice varying from the law in the books. Whilst the new rules were heralded as a radical departure and significant advance, they were never quite so radical as they were claimed to be. One of their main aims was to expedite trial, eliminating the so-called battle of experts by limiting the issues in contention at trial. In this way time and money would be saved. Lord Denning called the move 'altogether desirable in the search for justice and the saving of expense'. Others claimed that the new rules would put an end to behind-the-scenes manipulation of facts. The courts would be left with those real matters of controversy about which experts could not agree. The presumption that greater openness would automatically lead to greater consensus and reduced conflict amongst men of science was never addressed.

9 Processing the Legal Story: State versus Accused

In this chapter I examine the work of forensic pathologists and forensic scientists in order to show how it is both structured by, and structures, the process of conviction. Much of what pathologists and forensic scientists do, how they do it, and what they make of it flows from cognitive and interpretative commitments embedded in the process of prosecution. These may be found at the very start of a criminal investigation, in the processes employed at the scene of the crime to extract the information necessary to construct a case.

Pathology, Forensic Science, and the Police

A common misconception is that pathologists are part of the Home Office Forensic Science Service. Pathologists are in fact structurally separate from the HOFSS, though they were indeed the forerunners of State forensic science and the first laboratory at Hendon was headed by Dr James Davidson, a pathologist who had formerly worked as a lecturer in forensic medicine at Edinburgh. Despite this link, forensic pathologists never formally became part of the HOFSS structure. They were, and are, medical men generally working in hospitals and universities. When they investigate sudden and suspicious deaths they do so officially on behalf of the coroner. This is seen as a guarantee of their independence from the police. As this chapter will show, however, this independence is extremely fragile. Pathologists have a foot in two camps: they work as professionals in their own field but also as servants of the State. Both these footings would seem to guarantee the pathologist's claim to altruistic professionalism. Paradoxically, however, the pathologist's association with the State may come to undermine his claim to altruism: he may be seen as a hired gun of the State rather than as an independent professional person serving the public good.

The association between forensic pathology and the State is never very far from the surface, as the history of forensic pathology shows. Although pathologists and forensic scientists differentiate their disciplines, they are closely linked historically. What began as a semi-formal link in the nineteenth century was consolidated in the 1930s, when Lord Trenchard sought the assistance of pathologists in building up an organized forensic science service as part of a broader network of crime control. The early Hendon laboratory was attached to the Hendon Police College, where staff of the laboratory contributed to police training. Forensic science was described as 'a new tool which the police had not yet learned how to use, and did not quite trust'.[1] Initially consisting of only eight staff, it experienced rapid expansion in the post-war period, after the laboratory had been moved by Sir Harry Scott to Scotland Yard, where it became the Metropolitan Police Laboratory. The small local laboratories set up in the 1930s to serve local police forces (and the Metropolitan Laboratory set up to help the Metropolitan Police Force) still provide the organizational framework for modern forensic science. In the 1940s and 1950s they provided—along with Interpol in Europe and the FBI in America—part of the apparatus for controlling the enemy without as well as the enemy within.

Pathologists: Structural Links with Prosecution

As we have seen, towards the end of the nineteenth century pathologists were co-opted by the State to assist in the detection and conviction of offenders. Experts were retained by the Treasury Solicitor to appear for the Crown. For most of the twentieth century, pathologists have remained a small and élite circle of experts linked to the Crown by a number of formal structures. Foremost amongst these links is the association between pathologists and the Home Office. Pathologists are

[1] H. J. Walls, *Expert Witness* (John Long, London, 1972), 16.

retained by the Home Office, they are appointed as consultants to one of the Home Office laboratories, and they service the police. They are appointed after consultation with the Royal College of Pathologists and the director of the local Home Office Forensic Science Service. The hospital employing the pathologist receives a small retainer from the Home Office; he himself receives a call-out fee directly from the police. Fees for a specialized post-mortem are paid by the Home Office.

Historically, the exception to this system has been London, where pathologists are said to have greater independence from the police and are allowed to act for the defence:

> They are the source of the defence in criminal cases. Whilst one cannot criticise the forensic scientists in the forensic science laboratory service they are, whether they like it or not, principally working for the prosecution. Many of the legal profession prefer not to use their facilities but to use the facilities of academic departments such as serology for disputed paternity and stain technology and possibly for photography and also for toxicology.[2]

However, in 1988 the Home Office proposed to establish a Home Office list in London and set in train a system whereby the police would pay directly for the services of forensic pathologists.

There are a number of formal links between the State and forensic pathology. In a letter to the Clerk of the Home Affairs Committee in 1988, Professor Bernard Knight, Professor of Forensic Pathology, University of Wales College of Medicine, underlined the police–forensic science–forensic pathology link:

> I feel that as one of the 'forensic sciences', forensic pathology cannot be divorced from the service which is given to the police in the investigation of crime. As you will know, forensic pathologists work very closely with the Forensic Science Service, acting as a team at scenes of crime and have very close liaison in relation to the provision of scientific material for laboratory study, the interpretation of the results, especially in toxicology and in the presentation of evidence at trial.[3]

[2] HAC Report on *The Forensic Science Service* (HMSO, London, 1989), vol. ii, Minutes of Evidence, Sir Bernard Knight, 156.

[3] HAC Report (1989), vol. i, p. xi.

This association between the police and the pathologists structures their work. As Sanders has argued, cases are not simply sets of objective facts which can be ascertained once and for all. 'Cases have to be built out of what people say or think about the facts.'[4] He goes on to say that a number of different structures of fact could arise from an incident: 'What this means is that within an incident there could lie more than one kind of case. A case has to be selected by the police and choice of a particular case may determine the particular facts which are selected and presented.'[5] This is exactly what happens when the police and the pathologist attend the scene of crime. Pathologists are usually called to the scene of crime directly by the police, and whilst carrying out work at the *locus* they are surrounded by police officers. Working alongside the police they become imbued with police culture. They each create the other's interpretative context for action. The experienced pathologist links together certain kinds of information, and 'constructs inferences about the relationships among the surrounding elements'.[6] Courtroom stories are built up in this way into representations of reality, a story for courtroom production:

> Stories are important in this context because they are capsule versions of reality. They pick up an incident and set it down again in another social context. In this transition, data can be selected, the historical frame of reference can be specified, the situational factors can be redefined, and 'missing observations' can be implied. In short, a situation can be re-presented in a form consistent with an actor's perspective and interests both during and after the incident.[7]

Unlike most other types of forensic expert pathologists are involved in an investigation right from the beginning. Indeed, what the pathologist finds often determines whether there is a criminal case at all, or, as Professor Keith Simpson puts it,

[4] A. Sanders, 'Constructing the Case for the Prosecution', *Journal of Law and Society*, 14/2 (Summer 1987), 229.

[5] Ibid. 228.

[6] W. L. Bennett and M. S. Feldman, *Reconstructing Reality in the Courtroom* (Tavistock Press, London, 1981), 41.

[7] Ibid. 37.

'Nine times out of ten it is the pathologist who is able to say whether a crime has been committed or not.'[8] The pathologist is thus a primary player in the organization of the scene of crime. Together with the police, he interrogates the scene and his reconstruction of the dead defines the parameters of the case and determines what evidence should be sought, how it should be sought, and how it should be interpreted when it is found. The evidence which the pathologist finds does not 'miraculously appear after an incident ready formed like tablets of stone on Mount Sinai'.[9] Typifications of an incident based upon past experience provide him with a set of preconceptions from which to judge the meaning of the scene. Such typifications themselves eliminate ambiguities and omit some material. The interrogation of the scene is guided by what the investigators have come to regard as relevant evidence. They do not extract from the scene all that it may yield, but only that which is needed in order to construct a particular case. Experience over time means that both the police and the pathologist learn to construct cases more effectively. Thus begins the unidirectionality of the process of investigation.

However, experienced pathologists are generally much more reluctant than their younger colleagues to give definitive answers to questions about time of death, cause of death, and so on. Over the years they have come to realize that the physical evidence alone cannot answer these questions. Answers can only be framed in terms of an interpretation, framed in some context. Pathology itself has developed as a science mainly as a result of the need to provide proof that a suspect was in fact the offender. Shorthand categories have been constructed to aid this goal by filtering in (as well as out) certain types of physical information. Simpson's work on dentition matching provides an example of how such categories are developed. His interpretation of marks on the skin of a victim as bite marks provided a new means of identifying suspects. The interpretation of such marks has since featured in a num-

[8] *Sunday Times*, 8 Sept. 1984.
[9] D. McBarnet, *Conviction* (Macmillan, Oxford, 1981), 100.

ber of other cases. At the inquest into the death of Helen Smith, the pathologists' evidence varied significantly. Were marks of bruising on her thighs the result of forcible rape or her alleged fall? One witness, Dr Green, reported a fracture to the right shoulder which no other pathologist could see. He also mistook important facial bruising for henna dye from the victim's hair. In a similar controversy in Australia, it was alleged by an eminent prosecution expert (Professor Cameron) that marks found on some baby clothes indicated a bloody hand print: '[There was] the impression of a small adult hand in transferred blood, with four fingers at the back and the thumb at the front, as if the baby had been held by a hand under her armpit.'[10] There was a clear inference that the mother, Lindy Chamberlain, had literally had a hand in her baby's death. Professor Cameron examined clothing from the baby using ultraviolet fluorescence as an aid to distinguishing blood from other stains. He concluded that the neck of the baby's jumpsuit had not been cut by dingo teeth but by a knife or scissors:

> As to the possible causes of death, in the absence of a body, one must assume an unascertainable cause of death. Having said that, however, in the presence of the bleeding jump suit and from its amount and various other findings at that moment in time, it would be reasonable to assume that she met her death by unnatural causes, and that the mode of death had been caused by a cutting instrument, possibly encircling the neck, certainly cutting the vital blood vessels.[11]

The bloody hand print was said to have been caused by someone whose hand 'was wet with blood: a print consistent with a female hand'.[12] Other scientific witnesses found it difficult to see the bloody hand print at all. At trial Cameron was examined about his role in the *Confait* case, where he agreed that he had reached his conclusions without knowing all the circumstances of the case. The defence demonstrated

[10] I. R. Freckleton, *The Trial of the Expert* (Oxford University Press, Melbourne, 1987), 161.

[11] K. Crispin, *The Dingo Baby Case* (Lion Publishing, Tring, 1987), 80.

[12] Ibid. 84.

that his evidence in the *Chamberlain* case had also been based on false assumptions. Later investigations strongly supported the view that the bloody hand print was in fact the pattern caused by red sand percolating through the pattern of the baby's matinee jacket. They also showed that 80 per cent of the staining on the baby's jumpsuit was caused not by blood at all but by ferric oxide (red sand). Copper dust was also found to give the same test results as those relied upon by the prosecution to establish the presence of fetal blood. The Chamberlains lived in a district where copper dust abounded. The final inquiry decided that the spray of blood was a spray of sound-deadening material and/or ferric oxide.

Other scientific witnesses testified that stains found inside the Chamberlains' car were of fetal blood. The spray pattern of the stains inside the car was consistent, the Crown argued, with Lindy Chamberlain having cut the baby's throat in the front seat of the car. The scientific finding of fetal blood was contested at trial but the prosecution expert evidence prevailed. At a later inquiry, it was found that a sound-deadening material (Dulux Dufin 1081) used by the manufacturers in the production of Holden Torana motor cars produced the same spray pattern as that found inside the Chamberlains' car. It was found on large numbers of the same model throughout Australia. The argument that the spray of blood was the result of arterial blood spurting upwards under the dashboard disintegrated. To the layperson it seems astonishing that the experts could mistake copper dust and sound-deadening material for bloodstains. Evidence about blood is relied upon every day by the prosecution to obtain convictions. It is generally regarded as conclusive, a good example of the facts speaking for themselves. The *Chamberlain* case demonstrates how context structures the interpretative framework of forensic investigations. Pathologists construct categories by calling out of an incident or item particular interpretations of the facts which themselves then become factual. Just as the trial lawyer orchestrates his evidence to present an unambiguous case, so the pathologist and the forensic scientist reconstruct a disputed incident, mark, or sign to construct an unambiguous version of reality. Indexes of fingerprints, hand prints, bite marks, footwear impressions, blood groups, and DNA profiles are all typifications which structure future interpretations. They are also one way in which forensic pathologists and scientists contribute to a growing system of identification and surveillance of both the living and the dead. They reconstruct clear versions out of muddled events so that the reconstruction which appears at trial will often be clearer than the event was in its natural state.[13] As the *Confait* case will show, the dead may also be reassembled in particular ways in order to achieve convictions against the living.

Forensic Scientists: Structural Links with the Prosecution

Forensic scientists differ from pathologists in that they have a formally established niche in the structure of prosecution. They are endorsed by the State, they enjoy job security and high status as expert witnesses. In the sense that they are State servants, we might expect forensic scientists to be even more closely associated with the altruistic service motif than forensic pathologists. Indeed, this identification with public service has supported their image as impartial men of science contributing in a purely altruistic fashion to the public good. By contrast, the free-lance forensic scientist must always appear as the hired gun, a role which immediately threatens his claim to professional altruism. It is therefore all the more paradoxical that whilst State forensic scientists can make a strong claim to this service motif so distinctive of the professional ethos, the price which they have paid is the loss of their professional autonomy. They are civil servants and part of the Home Office Forensic Science Service, answerable not to equal professionals but to a hierarchy of scientists, managers, and bureaucrats. They are, therefore, very much pinned into an organizational structure over which they themselves exercise relatively little control.

[13] Bennett and Feldman, *Reconstructing Reality*, 167.

These formal links with the State were (and are) consolidated by the fact that HOFSS scientists work almost exclusively for the Home Office and the police. Within the general framework of the civil service, the HOFSS is managed by a mixture of civil servants and police and prosecution representatives. It is headed by a director-general, a Grade 4 civil servant, who reports to the head of the Science and Technology Group of the Police Department; he or she is, in turn, answerable to the Home Office. Incumbents of both these posts are assisted by a Policy Advisory Board consisting of several eminent scientists, the Chief Constable of each HOFSS area, an independent member with experience of the private sector, Home Office representatives, and a representative of the Crown Prosecution Service. This Policy Advisory Board decides the range and quality of services offered by the HOFSS. Its main function appears to be to manage and develop the HOFSS 'with a view to ensuring that manpower and financial resources meet operational needs efficiently, economically and effectively'.[14]

The failure of forensic scientists to establish themselves as a profession independent of the State means that there is really no private sector body to which defendants in the criminal justice process can turn for advice. In the mid-1980s this position began to change as some HOFSS scientists left the civil service in search of better pay, improved career prospects, and increased professional autonomy. A small independent profession began to form. Former HOFSS scientists set up their own forensic science consultancies, others moved into the pharmaceutical industry: 'Some scientists have left to work in private practice and more are planning to do so. Many say that if they were given a lead, and the terms looked attractive, they would follow.'[15] These scientists proved a new resource for the defence. They broke the Crown's monopoly on expertise. In 1988 the staff at Aldermaston and the CRE laboratories began to complain that their job was being made increasingly difficult 'as highly paid defence experts are often encountered [and]

[14] HAC Report (1989) vol. i, p. x.
[15] HAC Report (1989), vol. ii, appendices, 224.

reporting officers [are] frequently being exposed to lengthy and aggressive cross-examination in court'.[16] The Forensic Science Society made the same point: 'Many scientists who advise the defence are ex-Home Office scientists, thus the prosecution witness cannot hide behind any mystique of technical infallibility which may at one time have prevailed. This is healthy and any scientist not wishing to take part in the rough and tumble of our adversarial legal system is probably better out of it.'[17] This exposure to detailed and authoritative cross-examination has been a novel experience for State experts, unused to having their evidence challenged by persons equally familiar with the field of expertise and the in-house practices of the HOFSS.

The movement of forensic scientists out of the HOFSS must be related to low pay and falling morale within the service after a series of harsh criticisms and management reviews in the 1980s. The decline began in 1981 when Dr Alan Clift, a senior forensic scientist working for the HOFSS, was criticized by the Scottish High Court of Justiciary for the part he had played in the conviction of John Preece in 1973. Dr Clift's fall from grace was due to his supposed conviction-mindedness. The view espoused by the High Court of Justiciary was that scientists were supposed to be impartial and objective; their role was not to help secure convictions. This was also reiterated on behalf of the HOFSS by Dr Margaret Pereira.[18] she concluded:

> In many ways Dr Clift's attitudes reflect those of the very early forensic scientists who saw their function as one of 'helping the police' and not as I believe a modern forensic scientist would see it (a) to assist police in their investigations and (b) to assist in the cause of justice in the courts. He does not seem to have turned his mind to the possibilities of his evidence incriminating innocent people, trusting that the police were always right in their initial suspicions.[19]

In the fall-out from *Preece* the HOFSS rapidly sought a new image for its scientists. It made much of the theoretical availability of HOFSS laboratories to the defence and fought hard

[16] HAC Report (1989), vol. i, p. vii.
[17] Ibid., p. x.
[18] *Guardian*, 27 Jan. 1984.
[19] *New Scientist*, 3 Sept. 1981.

against the idea that it was simply an arm of the prosecution. However, even Dr Pereira was forced to acknowledge that the Home Office laboratories were never intended to be a facility for the defence. They were to provide a means of improving police evidence.[20] This was exactly the idea which the HOFSS now sought to remove by stressing that HOFSS operatives had no interest in conviction or acquittal. Dr Clift provided a very convenient opportunity to cast off the old image and put on the new. But this new image was undercut by the continued existence of strong structural links with the police. These were apparent in the management structure of individual HOFSS laboratories and in the Service as a whole, in the connection between the HOFSS and the civil service, and in the very structure and organization of HOFSS science itself.

The view that forensic science is and ought to be an arm of prosecution was in fact reiterated by the Home Office in 1988. The main function of the HOFSS was to provide impartial information about physical evidence recovered from the scenes of crimes.[21] Nowhere in law or practice is the role of forensic scientists laid down as anything else. In fact, forty-one police forces in England and Wales are served by the HOFSS, with the Metropolitan Police being served by its own Metropolitan Police Forensic Science Laboratory.

How exactly is the HOFSS linked to the police and prosecution? In 1988, the Home Office stated that the main objectives of the Forensic Science Service were 'to provide scientific resources to assist the police in the investigation and detection of crimes and to assist the courts in the administration of justice'.[22] Clearly, then, the Home Office does not regard the HOFSS as being independent of the police. Secondly, the HOFSS was traditionally funded from the vote for Police Support Services. Some costs were met by the police authorities from the Common Police Services Charge. As we shall see, this system came under scrutiny since it did not distinguish between costs incurred by individual police forces. This led to proposals in the 1980s for a restructuring of finances which

[20] Minutes of the Meeting of the Public Defender Committee, 13 Feb. 1989, 10.
[21] HAC Report (1989), ii. 10.
[22] HAC Report (1989), vol. i, p. xi.

placed the burden of costs directly upon users of the service. When this issue of costs was addressed by the Home Affairs Committee, it commented on the need to strike a balance between 'a service which could deal with all the cases to which forensic science could be applied and a service which is able only to deal with cases where forensic science will directly affect the outcome'.[23] The Rayner Scrutiny in 1981 recommended that the level of the service should remain relatively constant 'with greater emphasis on those cases where forensic science could offer most'.[24]

Apart from being part of the Home Office, the HOFSS is also part of the civil service. The 400 or so staff of the HOFSS were civil servants working in six laboratories: Aldermaston, Chepstow, Chorley, Huntingdon, Wetherby, and Birmingham, Home Office policy in the 1970s being to close down the smaller local laboratories and move towards economies of scale. It has been said that this move isolated the HOFSS scientists from the police and made them less vulnerable to the accusation of 'prosecution-mindedness'.[25] At the national level, however, the HOFSS retained close structural ties with the police. The break was not popular with local forces, who complained that the revamped HOFSS was failing to meet local policing priorities. In the 1980s the view which came to dominate the debate about the future of the HOFSS was that it should revert to its former self, and become more responsive to local policing. In paragraph 9 of its Report, the Home Affairs Committee specifically pointed out that increasing police dissatisfaction with the HOFSS was one of the main criteria for reassessing the future of the Service. In paragraph 27 it commented that a number of factors combined to 'create difficulties in the relationship between the Forensic Science Service and its principal customer, the police'.[26] It went on to say that since the preponderant customer for HOFSS services was the police, it was a matter for concern when the Association of Chief Police Officers claimed that the Forensic Science Service

[23] Ibid.

[24] HAC Report (1989), vol. ii, appendices, 224.

[25] HAC Report (1989), vol. i, p. vii.

[26] Ibid., p. xvii.

in England and Wales was unable to meet the demands of the police service.

The fourth structural link between the HOFSS and the police may be found in the organization of User Boards. Since 1987, each laboratory has had a User Board made up predominantly of police officers. It includes the Chief Constable of the area, a Crown Prosecution Office representative, representatives of other police forces in the area, and a representative of HM Inspectorate of Constabulary. Chief Constables who are members of local User Boards are also members of the National Policy Advisory Board. According to the Home Affairs Committee, these User Boards 'tighten links with the police service and make the service more responsive so that they can serve the police more effectively'.[27] Defence representation appears to be non-existent both on local User Boards and on the National Policy Advisory Board.

Apart from these structural links, there are other less tangible links between forensic science and the prosecution. These manifest themselves in the selections made in the course of scenes of crime investigations, in the typifications about cases and the kinds of evidence one requires to establish a case. They are replicated in the types of equipment scientists routinely take with them to a given scene of crime and the kind of data they collect there. In the 1970s and 1980s, senior HOFSS staff had noticed a large rise in exhibits submitted to laboratories. This was due, it was argued, to the use of non-scientific staff at scenes of crime, which would

> undoubtedly lead to a large increase in exhibits, many of which would not be of value and under normal circumstances would not have to be examined. The use of scene of crime officers where previously senior scientists attended major incidents means that a police officer says to a scene of crime officer 'you will take everything', results in just that, and the early sieving of items which would be likely to lead to information valuable to the police and a subsequent court has thereby ceased.[28]

[27] HAC Report (1989), vol. i, p. xvii.
[28] HAC Report (1989), vol. ii, Minutes of Evidence, 167.

Scientists have also been encouraged to attend the scene of crime 'to see in situ what is the situation and to advise on samples and exhibits that should be taken . . . There are some cases . . . where in practice it is by far the best for the scientist to work actually at the scene himself selecting the exhibits that it is necessary to take.'[29]

Clearly, then, selections are required from the very beginning of an investigation. The crime scene must be managed to produce evidence and exhibits: 'You must get the sequence right so as to get everything out of that scene or as much as you possibly can.'[30] Forensic experts examine what has been collected in the laboratory. They are aided in this by an account of the case furnished by the police by way of the HOLAB form. Where instructions to the HOFSS laboratory are vague, decisions about what to look for and which tests to run must be taken by the forensic scientist himself. He does this using prior typifications of similar cases. Gaps in the information given to the HOFSS scientist may not, however, simply be the consequence of sloppy police instructions. A directive by the DPP in the early 1980s apparently ruled that HOFSS scientists should not be given access to the witness statements in a particular case 'because it might make them more biased'.[31] The completeness of the information supplied to HOFSS scientists may thus be marred by police selectivity, poor instructions, and quasi-legal strictures. These may mean that the HOFSS scientist has only partial information upon which to found his conclusions:

> This makes the job of the forensic scientist very difficult, because increasingly these days, when he is expected to select out which items in the case should be looked at on the basis of very limited information, it is a very difficult thing to do. I think scientists in general terms think that far from making them less biased, not reading the statements, it probably makes them more biased because all they have to go on is a short summary, sometimes perhaps only three or four sentences long, written by the police officer in the case, which accompanies the exhibits to the laboratory. It was a very major problem for the scientist, to suddenly be cut off from the witness

[29] Ibid. 137. [30] Ibid. 138. [31] Ibid., Dr Gallop, 87.

statements . . . At worst, it might very well be that if a scientist rang the police officer in the case to ask for more information . . . and said 'Please can you send me this witness statement, I need it', he might say to you, 'Sorry, I cannot do that. You might be biased'. If you are lucky, he might read you out the relevant passage. Many scientists felt this was no way, no basis to it, to conduct their investigations.[32]

This former HOFSS scientist goes on to warn that

Simply reading a summary prepared by a police officer is not always sufficient to enable the scientist properly to decide what items he must examine and which he can safely ignore. As matters stand, he is obliged to be ever more selective to cope with increasing workloads and demands for reduction in case turn-around time. When such selectivity is practised in the light of relatively little information, there is the danger that points of contention will not be properly addressed.[33]

Clearly, then, forensic scientists are well aware that selections take place and that they themselves are required to make them. The selections made are mediated by a number of factors, including the cost-effectiveness of running particular types of tests. In a summary case, for example, it may be deemed inefficient or uneconomic to run a particularly expensive scientific test, or to run a less expensive one too many times. The Home Office itself is quite clear about this: 'The FSS has since stopped analysing blood unless there is a suspect or the offence is serious. . . . Conventional bloodgrouping can provide only supporting evidence against a suspect. It does not appear worthwhile to examine routinely blood from scenes because of the limited return obtained.'[34] Where a laboratory is concerned with through-put of cases (for example, where a monthly league table reveals slow or fast laboratories) it may be disinclined to run lengthy tests. Tests which produce adequate results in the short term may be preferred to more time-consuming ones which are less cost-effective. The extent to which tests are run, and what types of test are run, what

[32] HAC Report (1989), vol. ii, Minutes of Evidence, 167.
[33] Ibid. 83.
[34] Ibid. 196.

instruments are used, and so on, are all therefore shaped by a number of imperatives, the most important of which are extra-scientific. The competence with which tests are carried out may also vary according to considerations such as costs, time, and the demands on the time of skilled staff.

Figures for 1987 showed a total of 437 scientific staff within the HOFSS laboratories. These staff tend generally to be chemists or biologists. Their work involves examination of a wide range of materials including paint, glass, textile fibres, hair, firearms discharge residues, explosives traces, fire accelerants, blood and bodily fluids, botanical material, weapons, vehicle components, identification of drugs and noxious substances, the analysis of blood, urine, and bodily organs. This nation-wide apparatus creates the impression that all crimes everywhere are soluble using the combined resources of the HOFSS: 'Whether it's a pellet from the leg of a Bulgarian broadcaster or a fragment of the Brighton bomb, under the microscope of the Metropolitan Police laboratory the tiniest shred of evidence can help solve the most monstrous of crimes.'[35] Yet figures for the HOFSS show that the six laboratories only handle about 20,000 cases per year, excluding analysis of drugs and blood/alcohol in drink driving cases.[36] HOFSS work is employed in less than 1 per cent of all police cases.[37] An article in 1985 estimated that only 3 per cent of all London's crimes were sent to the Metropolitan Police Laboratory. Interestingly, in his 1972 biography, Walls records a Committee set up 'over thirty years ago [which] . . . guessed that forensic science might profitably be used in about 4 per cent of recorded offences'.[38] Figures from the Home Office for 1978–87 show that the total number of cases submitted to all six HOFSS laboratories has dropped from 96,024 in 1978 to 54,934 in 1987. Overall, the number of cases in this period dropped by 43.6 per cent. The largest fall occurred between 1982 and 1983 and, according to the Home Office, reflected

[35] *The Times*, 7 Apr. 1985.

[36] Home Office Research Report, *The Effectiveness of the Forensic Science Service* (HMSO, London, 1986) (Ramsay Report).

[37] Ibid.

[38] Walls, *Expert Witness*, 196.

changes in analysis of cases of driving under the influence of alcohol. Between 1983 and 1987, the number of cases submitted fell by 27.7 per cent. Some laboratories (Birmingham, Chepstow, Chorley, and Wetherby, for example) have seen their case through-put fall by about 50 per cent in the ten-year period, although the number of exhibits submitted only fell by 1.4 per cent, and in Birmingham it rose from 57,263 in 1978 to 89,491 in 1987.[39]

One reason for this relatively low level of use of forensic science lies in the fact that most recorded crime is perceived as too trivial to warrant the costs of using forensic science to help solve it. Forensic science tends not to be used in the hundreds of low-level routine cases processed by the police each year. Biology cases, involving grievous bodily harm, murder, robbery, and sexual offences, have by far the highest average number of exhibits submitted.[40] This may help explain why it is that biology cases show the highest average costs per case as compared with cases in other areas such as chemistry, toxicology/drugs, documents, firearms, and driving under the influence of alcohol.[41] Biology cases form an increasing proportion of the workload of the HOFSS. This is further confirmed by Ramsay's finding that cases sent to the forensic science laboratories by the police, though few in number, tend to be high in seriousness.[42] Proportionately, the number of cases in which the HOFSS is involved is low—though it varies with levels of arrest and levels of use of forensic laboratories in different police force areas.

Ramsay's study showed that almost one-third of all cases in the sample involved some kind of physical injury. It also showed that in four out of five cases the suspect had already been established. In these cases, the police wished to confirm or strengthen evidence against a suspect:

> The most frequent contribution [of forensic science] was in providing supporting evidence for the prosecution in cases where a suspect had already been charged (in 76% of such cases) or, less often, in cases

[39] HAC Report (1989), vol. ii, Minutes of Evidence.
[40] Ibid. 11, 12, 13; annexes 4, 5, 6.
[41] Ibid., annex 14.
[42] Ramsay Report (1986).

where no charges had yet been brought (in 39% of these cases). In a small but important group, accounting for 5% of the total, the FSS fully cleared someone who had either been charged or else was a strong suspect.[43]

A high proportion of these suspects had previous convictions.[44] Forensic science thus contributes less to catching criminals than to the assembly of evidence against known perpetrators. Its image as an aid to detection has been displaced by its role as an arm of conviction. In the majority of cases of murder and sexual assault, for example, offenders are already known to their victims. As Michael Zander has pointed out, 'It appears that the police for the most part solve only the crimes in which the accused is identifiable from the outset, which obviously represent only a small proportion of all crimes. Moreover, they probably do not include most of the crimes committed by total strangers which cause the public most anxiety.'[45]

The tendency to employ HOFSS work only in serious cases where there is a known suspect arises in part from a policy of selectivity adopted by the HOFSS in the 1980s. In the post-Rayner era, HOFSS laboratories became much concerned with the rate of case turnover. Additional pressure was brought to bear by a provision of the 1984 Police and Criminal Evidence Act (PACE) which required fuller evidence, including expert evidence, at the committal stage. The HOFSS geared itself up to meet case turn-around times in the courts, rather than to meeting police and crime control requirements. The effect of this policy of selectivity was to reinforce the idea that forensic work should only be undertaken in cases where there was a known suspect. The police were apparently unhappy at the prospect of having cases turned away from the HOFSS laboratories, arguing that this discouraged them from sending some sorts of cases in the first place. However, the introduction of User Boards in the late 1980s appears to have been an attempt to provide a forum for discussion between the police

[43] Ibid. [44] Ibid.

[45] M. Zander, 'An Investigation of Crime: A Study of Cases Tried at the Old Bailey', *Criminal Law Review* (1979), 203.

and the laboratories about the policy of selection. The net effect was that 'The police know very well the kind of things they are expected to put forward . . . it will simply . . . enable the police force to reflect a little more easily as to which cases they want to pursue.'[46]

The combined effect of selectivity and cost-effectiveness is that forensic science is brought to bear mainly in order to assemble a case against a known suspect rather than to sift out cunningly clever criminals from the population at large. However, the perceived seriousness and public significance of these relatively few cases continues to give forensic science its high profile as the key to the State's successful detection of difficult and dangerous criminals.

Formal Cognitive Commitments to the Construction of Crown Cases

The process of prosecution also gives rise to particular sorts of scientific problems which scientists in other kinds of fields do not normally address. These special problems are dealt with by a Central Research and Support Establishment at Aldermaston, which is charged with the task of developing the scientific techniques used by the HOFSS. The dynamic for CRSE development of new techniques is supplied by casework brought to their attention by the police. The Home Affairs Committee is fairly explicit in describing the primary focus of CRSE work as the operational requirements of the police. Once a decision has been taken on whether or not to use a new scientific technique in casework, the necessary equipment is purchased for use in the laboratories and new training programmes are arranged to instruct operatives in the new techniques. The CRSE system is a good example of how cognitive commitments towards prosecution have become institutionalized. The operational priorities of the police and prosecution are entrenched in the material and intellectual resources (equipment, techniques, ideas, processes) of the HOFSS. The

[46] HAC Report (1989), vol. ii, Minutes of Evidence, 127.

HOFSS is a highly structured facility. Its architecture is geared to achieving very particular ends. The scientific products which furnish prosecution and conviction are occasioned by the circumstances of their production. The choice of measuring instrument, tools, and materials are also pre-defined:

> The choice of a particular measurement device, a particular formulation of chemical composition, a specific temperature or of the timing of an experiment is a choice among alternative means and courses of action. These selections in turn can only be made with respect to other selections: they are based on translations into further selections, the so-called decision criteria.[47]

The selections which inform the expert's work are essentially based in the adversarial process. If these selections come to light, they can be the undoing of the expert, for 'selections can be called into question precisely because they are selections; that is, precisely because they involve the possibility of alternative selections. If scientific objects are selectively carved out from reality, they can be deconstructed by challenging the selections they incorporate. If scientific facts are fabricated in the sense that they are derived from decisions, they can be defabricated by imposing alternative decisions.'[48] Alternative selections within forensic science allow a different legal case to be constructed from the same basic event. Once the police choose their case, they are not interested in knowing about all likely interpretations, but the one which best fits. Selections which have previously proved successful will be made and will continue to set the scene at the site of action. They will, in short, not only be decision-impregnated but decision-impregnating.

A More Dedicated Tool of Investigation

At both the practical and the ideological level the forensic pathologist and the forensic scientist do a compendious job for the State. As Lindsay Prior argues, in so far as pathologists

[47] K. Knorr-Cetina, 'Towards a Constructivist Interpretation of Science', in K. Knorr-Cetina and M. Mulkay, *Science Observed* (Sage, London, 1983), 121.
[48] Ibid.

police the dead, they form part of society's regulatory, supervisory, and advisory powers:

> State pathologists can be routinely involved in the exercise of regulatory and observational power on behalf of the state for both repressive and productive ends, and the modern forensic pathologist is as concerned with threats to public health as he is with the details of criminal activity. In fact, the application of the principles of scientific pathology in the mortuary combined with the collection of 'vital' statistics in state bureaucracies form two great pillars of the contemporary apparatus of medical surveillance.[49]

The view of the Fisher Inquiry into the *Confait* case was not that pathologists should be more detached from the police but that they should be more in tune with police thinking. This is, of course, quite contrary to the rhetoric which portrays the man of science as an independent and impartial scientist, aloof from and uninterested in conviction. Despite the furore which followed the Clift cases in the early 1980s, a number of official bodies in fact supported the view that forensic experts should be closely tied into police work. For example, when the debate about the future role of forensic pathologists initiated by the Fisher Inquiry was taken up by the Home Office, it argued that forensic services should be paid for by the customer using them, that is, the police. There was also concern that the dwindling numbers of qualified pathologists threatened to leave the police without adequate forensic support. University cuts in the 1980s left university posts in forensic medicine unfilled; the ageing of the profession and low levels of recruitment led to speculation that there would, in years to come, be a shortfall in the number of forensic pathologists. New pathologists were said to be insufficiently aware of police needs. In his evidence to the Home Affairs Committee, Professor Bernard Knight noted some anecdotal evidence that, as the older senior pathologists were vanishing, new pathologists (some recruited from the NHS) being brought on to the Home Office lists 'were by no means finding favour with the police'.[50] One response of

[49] L. Prior, 'Policing the Dead: A Sociology of the Mortuary', *Sociology*, 21/3 (Aug. 1987), 355.

[50] HAC Report (1989), vol. ii, Minutes of Evidence, 157.

the British Association of Forensic Medicine was to ask the Royal College of Pathologists to re-evaluate its criteria for recommending new pathologists to the Home Office list. The argument for doing this is based on the following view: 'frankly some of the appointees to the Home Office lists from the Health service are too inexperienced to do the job. If you want a brain tumour operated on, do not go to a skin specialist. The same sort of thing is happening in our speciality: people who are not specialists in our trade think they can do it. Some of them are keen to do it, but that does not mean to say they are able to do it.'[51]

The problem, therefore, was not simply a shortage of pathologists but a shortage of pathologists sensitive to the needs of the police. The situation in the mid-1980s was as described in the following terms by one pathologist: 'I have not seen anything actually fall apart at the seams yet, but I suspect that it is only a matter of time before it does. We are living on borrowed time. It is working today but I suspect it might not work quite so well tomorrow.'[52] One remedy was to consolidate existing relationships by making the police pay directly for forensic services. A Home Office Working Party on Forensic Pathology (the Wasserman Committee) set up in 1984 recommended that forensic pathology be restructured so that services would be paid for directly by the police. This would also mean that pathologists were more responsive to police needs. The proposal worried pathologists, who believed that their independence, integrity, and impartiality would be compromised: 'We would probably become labelled "police" pathologists and subject to their orders and wishes, with all that could mean in the quality of evidence.'[53]

This restructuring of forensic pathology meant that the local customer rather than central government would pay for the service provided. As Dr Green pointed out in his evidence to the Home Affairs Committee, one consequence of this system might be the 'temptation to go cowboy', that is, employ cheaper and less qualified experts rather than the more

[51] Ibid. [52] Ibid., Dr Alan Green, 160. [53] *Guardian*, 3 July 1985.

expensive qualified experts. This concern was echoed by a Home Office memorandum to the Home Affairs Committee: 'if the police were in fact responsible for funding the Forensic Science Service they would probably opt for the cheapest deal and that quality would be sacrificed.'[54] The new funding structure would tie forensic services more closely into police priorities. This undercuts the prospect of an independent profession of forensic pathology equally available to all.

In the HOFSS, a similar change in procedure and funding structure was coupled with the reforms of the Police and Criminal Evidence Act of 1984. An equally important factor was central government's concern with the costs of the HOFSS. This concern was consolidated by the findings of the Rayner Review, where we first find full expression of the desirability of economy, efficiency, and effectiveness in public service agencies. In the 1980s, these three Es became the touchstone of HOFSS organization. To achieve the three Es HOFSS processes had to be broken down into units of production and unit costs. Assessment of costs was to be assisted by the introduction of time-recording systems, which allow work undertaken to be charged to a particular client or customer. The alliance between quality control and unit costing began to structure the HOFSS in very particular ways. For the scientists themselves, it brought a loss of individual discretion and status; client relationships were to be based on the cash nexus; cost-effectiveness demanded that scientists subordinate their work to a kind of technological managerialism, with a consequent loss of professional morale:

> In those days [twenty-four years ago] Forensic Science was a craft which one learned by long apprenticeship and practised with great pride in one's work, but this, I think, is no longer the case. Establishments have become bigger and more impersonal and the craft has become just a job, less and less well paid in relative terms and more demanding in terms of work load, outside interference and hassle from the top . . . Today's Forensic Scientist in the Home Office is expected to function like a robot in a car factory; all the necessary materials and resources are provided; the working environ-

[54] HAC Report (1989), vol. ii, Minutes of Evidence, Dr Alan Green, 163.

ment is well-maintained; the quality of the product is constantly checked. With such a system you get the right result every time and you don't have to pay robots too much either.[55]

Of course, the revamped image of the forensic scientist as an impersonal cog in the State machine is also something of a myth. It conjures up an image of faceless men and women going about their work in a wholly impersonal, standardized manner, more concerned with the bureaucratic aims of efficiency and effectiveness than with conviction. In practice, the materials for examination are still put into the system by the police. These may be very different from the samples put into the system during blind trials. The police may pre-select the materials which they wish to be forensically examined, and keep back other materials which they do not wish to be examined. The police also set the interpretative context for investigation:

> A scientist has material referred to him by an investigating officer. The investigating officer will, in the referral form, to a greater or lesser extent set out the parameters of the investigation and of the potential significance of that particular item as perceived by the referring officer. The scientist would examine the material within the context of that general definition of what is involved. On the basis of that, he would provide an opinion. If he is a very good scientist, he will look at the results and he will say: 'Well, you know, there are other explanations than those which are, as it were, sought for in this particular referral form'. If he is not, or if he is too busy, or if he does not have enough experience . . . he will confine himself to that which he has been asked to deal with. . . . the real problem we perceive is in the context of opinion and its commitment to a particular cause.[56]

It is police officers to whom the scientists must turn for more information, and it is to police officers that they must return their completed reports. In his evidence to the Home Affairs Committee, Henry Bland drew attention to the fact that quality assurance is only as good as the material received at the laboratory: '[It] does not extend to the collection of the

[55] Ibid., letter from David Jones, 120. [56] Ibid., Michael Hill, 150.

specimens for examination . . . a scientist can only examine the materials sent to him in the light of the questions posed. He accepts on trust that the integrity of the samples has been maintained throughout unless it is blatantly obvious that some error has occurred.'[57] The Criminal Bar Association noted that it matters not whether the withholding or contamination of information is deliberate: 'the consequences are potentially very serious.'[58] Whether the forensic scientist is aware or unaware that his investigations and opinions have been shaped by the terms of the referral, he will still develop the Crown's case against the accused in the required direction.[59] The CBA goes on to give examples of cases in which evidence was withheld which was of significance to the defence.[60] It recommends that the HOLAB form should be attached to the expert's report, so that the defence is able to see what assumptions underlie that report.

In its evidence to the Home Affairs Committee the Criminal Bar Association attempted to raise some of these issues. In reality, it argued: 'The only true independence which the FSS has from the police is the independence of establishment. It is our strong view that it is a fundamental mistake, both in fact and in the long term perspective, to confuse independence of establishment (or, even, theoretical independence) with independence of function or approach.'[61] The possibility of police contamination of materials is in fact used to justify a closer relationship for, it is said, only by closer liaison can unintentional physical contamination be prevented. Properly instructed scenes of crimes officers will be more aware of the possibility of contamination and take steps to avoid it. This more or less guarantees that the facts will be structured after a particular fashion. Quality assurance programmes have a capacity to disguise this structuring. The results of quality assurance tests are sent on to the CRSE, as are the reports from all the other laboratories. They are not made public. They are discussed internally. An additional system of internal auditing compounds this secrecy. It relies upon inspection of a laboratory every

[57] HAC Report (1989), vol. ii, Minutes of Evidence, Henry Bland, 168.
[58] Ibid., 166. [59] Ibid. 151. [60] Ibid. 146. [61] Ibid. 143.

three years by three assistant directors from other laboratories. The system is 'prone to collusion . . . it can easily turn into a case of "I won't criticise you this year because you may have the chance to criticise me next year".'[62] The system therefore lacks independence and public accountability. One is simply told that the integrity of those involved cannot be questioned. This is, quite literally, correct.

Justifying its refusal to disclose the outcome of quality assurance tests, the Home Office argues that these reports are technical matters, unsuitable for public consumption, 'and could not easily be adapted without over-simplifying the issues for the general reader'.[63] This has, of course, been the classic argument for self-policing adopted by the professions. At the same time, internal standardization encourages HOFSS scientists to produce reports which conform to the house style and which may, as a consequence, conceal any selections made in the course of forensic investigations. The effect of these two developments is compounded by the fact that in the 1980s HOFSS scientists were introduced to witness-skills training. This included practical sessions in the 'preparation of statements, followed by cross-examination by experienced criminal lawyers acting for the "prosecution" and the "defence"'.[64] The HOFSS recognized that impartiality is an image which must be accomplished. The expert: 'must make the court feel that he is a man of absolute integrity, that his opinions have been formed with scrupulous care and that every possible precaution to avoid error has been taken. He must further eschew unnecessary technical jargon and carefully avoid giving the impression that he is trying to blind them with science.'[65] As Home Office experts become better witnesses, it becomes even more difficult for defendants to puncture their image of infallibility. This is not to say that opinions given by HOFSS scientists are not honest but dishonest. But a polished witness performance makes it impossible to discriminate between the two.

[62] Ibid. 30–1. [63] Ibid. 196. [64] Ibid. 6. [65] Walls, *Expert Witness.*

Redressing the Balance? Independent Facilities

The State clearly enjoys a tremendous material advantage in the field of forensic expertise; it also enjoys a monopoly on credibility. Quite apart from the police laboratories and the six HOFSS laboratories, it also has at its disposal the facilities of the Ministry of Defence at centres such as Woolwich. It has 'most of the expertise and certainly most of the equipment within our own four walls. There are no defence experts who have anything like the facilities that we have in our laboratories.'[66] Moreover, since 'no-one has our equipment, one cannot get a second-opinion' The remedy for this inequality has been to offer selected defence experts ('not every Tom, Dick and Harry . . . but known experts') restricted access to HOFSS facilities, on the basis that if they use the State's resources they will always come to the same answers as the State. Differences between the two are explained away as differences of interpretation: 'If a finding is right there will never be any argument over the finding: it is the interpretation of the finding that is the important factor and that is really where the area of dispute between the two sides often lies.'[67] This view is clearly based on an understanding of science in which the facts speak for themselves. The defence expert has no opportunity to adopt a different point of view or use alternative techniques of observation. He must use only those means used by the Crown experts, he must use them in the same way as they have been used by the Crown experts, and he must apply them to the same materials as those used by the Crown experts. All this process allows the defence expert to do is confirm HOFSS findings. It puts a gloss on a particular set of findings.[68]

Alternative facilities for the defence have been mooted as one means of redressing this situation. The idea is not new; it has been aired in various forms for over half a century and it has gained the approval not only of defence lawyers themselves

[66] HAC Report (1989), vol. ii, Minutes of Evidence, 34.
[67] Ibid. 35.
[68] Ibid. 83.

but also of some members of the judiciary.[69] It is in precisely those cases where the defence has had access to other facilities that previously undisputed findings have been thrown into doubt. Alternative facilities also open up the possibility that the defence may supply its own exhibits for testing, particularly important where no exhibits have been supplied by the police.[70] The official reaction to the notion of independent defence facilities is that, since HOFSS scientists are already independent of the police, there is no need for any alteration to existing arrangements. The State's preferred option is for HOFSS laboratories to undertake all defence work as well as all prosecution work. This would undercut the market for independent experts in the private sector, leading in effect to State closure on forensic expertise. An authoritative critique of State forensic science is at present difficult; under these conditions it would become impossible.

The proposal to make the Forensic Science Service into this kind of independent executive agency took effect in 1991. In theory this new structure would remedy the disadvantaged position of the defence since, as an agency, the HOFSS could act for a variety of customers, including defendants. The HOFSS would only be allowed to take on work for customers other than the police once it had met the police demand for its services. Since police case work would retain priority, and since it is carried out in such high volume, it would easily swamp any work submitted by the defence. Moreover, since the police would be the biggest customer for an agency HOFSS, the net effect of this proposal would actually be to increase police control by making the HOFSS financially dependent on the police. The proposed changes therefore strengthen rather than weaken the cognitive and material ties between the prosecution and its experts. At the ideological level, however, the advantage of the agency concept is that it lends the HOFSS an air of independence. Several damaging

[69] Sir R. Ormrod, 'Evidence and Proof: Scientific and Legal', *Medicine Science Law*, 12 (1972); see also Sir R. Ormrod, 'Scientific Evidence in Court', *Criminal Law Review* (1968), 240–7.

[70] HAC Report (1989), vol. ii, Minutes of Evidence, Dr Brian Caddy, 71.

myths are therefore perpetuated by the revamped image of quality assurance and executive agency status. The first is that by formally distancing the HOFSS from the State the HOFSS will automatically become independent. In fact it becomes even more intricately tied into prosecution objectives. Ideologically, the image of the scientist as a mere cog in a vast impersonal machine also underlines this sense of impartiality. By obscuring the influence of the human hand upon scientific work, it contributes to a reification of science. Where science underpins legal decisions, it also endows them with a sense of immanence. Thus legal decisions come to be seen increasingly as the inevitable outcomes of the facts speaking for themselves rather than outcomes based upon choice and value.

In the construction of the Crown case, the police and prosecution are concerned with establishing not only facts about but also facts against the accused, that is, facts which tend to implicate him in a crime.[71] In the process, Sanders argues, the police do not find strong cases against the accused, 'They make them.'[72] They play down or eliminate all the features which tend to undermine the case against the accused, making cases 'as near a fait accompli as they can manage'.[73] Selections are made which eventually construct the case so that it is all one way—disconfirming factors are discounted and/or eliminated. It may not always be the strongest case or the most winnable, but it becomes so through the process of case construction. This is rather a different picture from the official version, in which one is encouraged to imagine a system of criminal justice 'which bends over backwards to favour the defendant rather than the prosecution'.[74]

In her book *Conviction* Doreen McBarnet asked how, given the ambiguities and uncertainties that dog real-life incidents, are 'clear cut facts of the case and strong cases produced'?[75] In the foregoing analysis I have tried to draw out the part forensic experts play in making the facts of the prosecution case become self-evident. The new emphasis upon standardization

[71] Sanders, 'Constructing the Case for the Prosecution', 245.
[72] Ibid.
[73] Ibid.
[74] McBarnet, *Conviction.*
[75] Ibid.

and routinization, on the impersonal nature of forensic practice, invites us to believe that if we get the science right, justice will follow. It also confirms the law's view that miscarriages of justice are due to bad science and bad scientists, rather than to bad law and bad lawyers.

10 The Impact of Advocacy

This chapter examines some concrete examples of how norms of advocacy and the permissions of the formal law have shaped specific cases. The material upon which I draw includes cases from my own fieldwork as well as criminal cases which in recent years have been subjected to intense judicial scrutiny. In some instances these sources overlap, since the fieldwork sample included cases which were subsequently reviewed either by the Court of Appeal or by judicial inquiry.

The *Confait* Case (1972) and The Fisher Inquiry (1977)

The *Confait* case is a good example of how, as McBarnet puts it, guilt has been organized out of the public eye; it shows us how 'Adversary investigation filters out ambiguities and leaves only black and white cases—caricature versions of reality offered by the prosecution and defence; courtroom powers and sanctions prevent any deviation in court from either of these filtered versions.'[1] Lawyers have a number of ways of making expert evidence support a black and white case. I have chosen the experience of Professor Cameron in the case of Maxwell Confait as one instance of this. It illustrates how, even when an expert is involved right from the start of an investigation, it is relatively easy to structure his input so that it favours the prosecution case. It also raises interesting issues about the proper role of the expert. Should he be a passive tool in the hands of the lawyer or is he under some wider duty to the court to take the initiative in bringing to its attention discrepancies in the evidence? Is there any duty on the expert to bring to the court's attention information which raises doubts about the police case and is material to the defence? These are

[1] D. McBarnet, *Conviction* (Macmillan, Oxford, 1981), 100.

important issues which are also raised by the experience of Dr Clift in the case of *Preece*.

Maxwell Confait's body was discovered on 22 April 1972 at 27 Doggett Road, Catford, south London. It was discovered by fire officers called to attend a fire at the same address. Two boys, Colin Lattimore and Ronald Leighton, were arrested for the murder of Maxwell Confait, and together with a third boy, Ahmet Salih, they were charged with setting fire to the house. Having arrested these three boys and charged two of them with the murder the police sought confirming evidence linking them to the crime. Unfortunately for the police their chief suspects had alibis for the time of death. Nevertheless, at trial Leighton was convicted of murder, Lattimore of manslaughter on the ground of diminished responsibility, and, together with Salih, they were convicted of arson.

The *Confait* case was to give rise to a number of misgivings. It raised the issue of confessions made by suggestible persons subjected to considerable psychological stress; it also raised important questions about the reliability of the forensic evidence. It was referred by the Home Secretary to the Court of Appeal in June 1975; in October 1975 the convictions were quashed. The Court concluded that the burden of proof had not been properly discharged and that as a result the convictions were unsafe and unsatisfactory. Why did the Crown case come apart? The Fisher Inquiry provides some answers.[2] Amongst other things, Fisher examined the role played by the forensic evidence. Time of death, notoriously difficult to assess, was crucial to the Crown's case against the suspects. For the time given by Professor Cameron, the suspects had alibis. The Crown consciously sought to stretch the period in which death could have occurred in order to neutralize this alibi defence. Sanders writes:

> The police knew that their case against one of the defendants (Lattimore) would fail unless his alibi could be neutralised. Lattimore's alibi covered the whole period in which the pathologist originally estimated Confait had died. The police therefore

[2] *Report of an Inquiry into the Death of Maxwell Confait* (Fisher Inquiry) (HMSO, London, 1977).

persuaded the pathologist to alter his estimate. The time of death was a fact unprovable; the next best fact was an expert's opinion as to when that fact had most probably been. The police case against Lattimore rested on the successful re-negotiation of the pathologist's estimate of Confait's time of death. In other words, the strength of the police case rested on the police's ability to transform one set of facts (the pathologist's opinion) in order to nullify the effect of another (Lattimore's alibi).[3]

Professor Cameron was called to the scene of the crime and examined the body at 3.45 a.m. He later estimated that, by that time, the man had been dead for some six hours plus or minus two hours. By the time he had started his post-mortem at 6.30 a.m. rigor mortis was complete, 'which suggested death 12 hours earlier'.[4] He was invited to a conference with the prosecuting lawyer Richard Du Cann two days into the trial. As far as he could recollect, he was asked about various factors which might have affected the normal onset of rigor mortis 'and whether it was possible to narrow the area over which he had estimated the time of death'.[5] On Cameron's own account, he had no idea why he was asked these questions. Their relevance to Du Cann was that factors which might have disturbed the onset of rigor mortis might also introduce a degree of uncertainty about the time of death. In opening the prosecution case Du Cann left the question of time of death wide open and called evidence to the effect that death 'could have occurred immediately before the onset of the fire, i.e. outside the periods mentioned in the two statements . . . At the conclusion of his case he was suggesting that death had occurred after midnight.'[6] This meant that the boys could still have committed the murder. Fisher concluded that 'the case was presented to the jury in a way which over-stated the uncertainty of the estimates of the time of death, and which over-rode the actual estimates which Dr Bain and Dr Cameron gave'.[7] The prosecution had, therefore, not only cen-

[3] A. Sanders, 'Constructing the Case for the Prosecution', *Journal of Law and Society*, 14/2 (Summer 1987), 240.
[4] Fisher Inquiry, 63. [5] Ibid. 64. [6] Ibid. 219. [7] Ibid. 64.

sored certain facts, it had also deliberately rendered some facts indeterminate in order to obtain a conviction.

Given that the prosecution had information which was material to the defence, one issue is, why was this not disclosed? What duties lie upon the professional officer of the Director of Public Prosecutions, to whom this case was allotted? In the *Confait* case, he saw his duty as being 'to decide upon the selection of witnesses, the editing of their statements if thought necessary, whether or not in appropriate cases the evidence should be served upon the defence . . . [and] to decide upon the order of witnesses and matters of that kind.'[8] It was his duty to see that there was evidence to support the charges. The DPP, Sir Norman Skelhorn, said that had his professional officer noticed discrepancies in the evidence, his duty would have been to have a conference with Professor Cameron to see whether his views could be reconciled with the confessions and the probable time of the fire. If the discrepancy had come to light after the committal stage, he should have drawn counsel's attention to it. The view taken by Sir Norman Skelhorn was that he was doubtful whether the discrepancy should have been spotted as important.

Fisher did not agree. He concluded that if what the professional officer of the DPP's department did was regarded as in keeping with prevailing practice, then prevailing practice was wrong. In his view, had the professional officer spotted the discrepancy and brought it to the attention of the magistrates at the committal stage, they might well have decided that there was no case for trial. Fisher also noted that had Professor Cameron been given sight of the other evidence, had he been asked to reconsider his evidence in the light of it, and had he been asked the relevant questions in a neutral way instead of being asked to suggest ways in which the period for the time of death could be extended after midnight, 'the course of the trial would have been very different and an acquittal might have resulted'.[9]

Like so many experts, however, Professor Cameron was not

[8] Ibid. 215. [9] Ibid. 223.

given sight of all the evidence. Building up a prosecution case is not simply a matter of eliminating all ambiguity. For some cases to succeed, it may become necessary to build in strategic ambiguities. The net effect of keeping Cameron out of the picture was to create sufficient ambiguity to gain a conviction. Fisher said that he was sure that the police had identified the expert's estimate of time of death as a difficulty and made no attempts to firm it up since this would have made their case more difficult to establish. Sanders writes that the police entered into some negotiations with Cameron over the time of death: 'The police then raised sufficient doubt in the pathologist's mind about the likely time of death to persuade him to "stretch" the possible period in which he (Confait) could have died, thereby nullifying the alibi. The logic of the police was faultless: if the suspects were guilty and the alibi could not be disproved then the alibi has to be inapplicable.'[10] The police did not disclose the fact that they had had negotiations with the pathologist to try and get him to stretch the time of death. They are under no lawful duty to do so. Neither were the actions of prosecuting counsel in any sense unlawful. Their approach was in keeping with the view that once the police had identified their chief suspects, their aim thereafter was to strengthening the case against them. To do this, they exploited legal procedures which allowed them to filter out damaging information. The facts of the case 'were created . . . as well as selected' in the context of advocacy.[11]

Pathologists generally consider that the best way to establish time of death is to take either the rectal or abdominal temperature of the body. This had not been done, Professor Cameron told Fisher, because he had some knowledge of the habits of the deceased. Confait was a known homosexual and was suspected of having engaged in homosexual intercourse shortly before death. Professor Cameron 'thought it would be wrong for him to disturb the body at the site, and that it would be better for him to wait until he had the body in more suitable surroundings'.[12] Cameron's knowledge of the back-

[10] Sanders, 'Constructing the Case for the Prosecution'.
[11] Ibid. 240.
[12] Fisher Inquiry, 70.

ground to the case, of the habits of the deceased, could only have come from the police at the scene of the crime, concerned to establish whether the death was a homosexual murder or gaybashing. An examination of the rectum might reveal evidence of homosexual intercourse. Textbooks on pathology advise that the pathologist should never take the rectal temperature in such circumstances, though the bodily temperature might be taken through an incision in the abdomen. The direction of Professor Cameron's examination was, therefore, highly structured. This directly influenced what evidence he looked for, what tests he carried out, and how he carried them out. Fisher clearly appreciated this. He stated that in practice they (the police and the pathologist) work as a team. In Fisher's view, problems arose in *Confait* not because this relationship was too close but because it was not close enough: the pathologist spent too little time considering issues of vital importance to the police. To remedy this, he set out a series of proposals which would tie the pathologist more closely into the police investigation. The pathologist should:

1. Be employed by the police and instructed to report to the police, though he might continue to work for the coroner as well.
2. Be told of any special matters of interest to the police and should deal with them in his report. A list of standard questions could be drawn up for the pathologist to answer.
3. Be given all relevant witness statements and be asked to reconsider his own report in the light of them.
4. Be in court to hear the other relevant evidence before he gives his own evidence.
5. Make every attempt to narrow and firm up estimates of time of death.
6. Always have a conference with counsel before trial.

The overall goal was to make sure that the pathologist clearly understood the needs of the prosecution.

The Dr Clift Cases, 1978–1984

Working with prosecution objectives in mind was, however, precisely the criticism levelled at Dr Alan Clift and several other Home Office forensic scientists during the 1980s. That decade saw a growing critique of forensic science, fuelled by public disquiet concerning a number of miscarriages of justice in which State forensic scientists seemed to have been prosecution-minded. The so-called Clift cases began in 1981 with an appeal case, *Preece* v. *HM Advocate*, in the High Court of Justiciary in Edinburgh.[13] The Crown's expert at the original trial, Dr Alan Clift, was branded 'Dr Blunder' by the Press and was said by the court to have been discredited as a scientist and as a witness. The court held that he had demonstrated 'a complete misunderstanding of the role of scientific witnesses in our courts and a lack of the essential qualities of accuracy and scientific objectivity which are normally to be taken for granted'.[14]

The facts of *Preece* are, briefly, that Preece was convicted in 1973 of murder by strangulation of a woman named Helen Wills. It was the Crown's case that he had killed her after or during sexual intercourse in the cabin of his lorry. Evidence of contact between Helen Wills, Preece, and the cabin was provided by the forensic scientists Dr Clift and Dr Gregory. The two experts produced a joint report in which they stated that they had examined the productions and found eighteen fibres in dust from the floor of the cab and from the sleeping bag and cushion found in the cab, which were similar to fibres found in the coat worn by Helen Wills. They also found a brown hair on her coat which differed from her own hair but was similar to that of Preece, semen stains on a vaginal swab from Helen Wills and similar stains on her tights and knickers, both of which indicated a blood group A secretor. Finally, a saliva swab from John Preece indicated that he was a blood group A secretor.

Because the body was originally found on the English side

[13] *Preece* v. *HM Advocate* (1981) Criminal Law Review 783–5.
[14] Ibid., per Emslie, LJG.

of the England–Scotland border, initial investigations began in England. Dr Clift, then Principal Scientific Officer at the Home Office North Western Science Laboratory, was asked to undertake the forensic investigation. Later, when it transpired that the murder itself must have taken place in Scotland, proceedings were moved north of the border. The shift in the location of the original investigation and trial proceedings is important. Dr Clift had prepared his original report for proceedings in England, where he was familiar with the legal process. The move to Scotland placed him in a completely different legal environment, where the law and pre-trial procedures differed. Arguably, this made him rather more dependent than usual on lawyers for advice and guidance in the case. In any event, he was asked to redraft his report in the style acceptable to Scottish courts. He did so. At trial, Dr Clift was asked by the prosecution about stains found on the tights and knickers of the victim. He gave evidence to the effect that these were mixed vaginal and semen stains, and that they indicated the presence of a blood group A secretor. Only about 30 per cent of the population have blood group A. The accused was not only of this blood group, he was also a secretor. Combined with the other evidence in the case, this was a major part of the Crown's case that Helen Wills had been in prolonged and close contact with John Preece. Preece was convicted.

The case was referred back to the High Court in 1981, under section 263 of the Criminal Procedure (Scotland) Act 1975. The referral grew out of an ongoing internal HOFSS investigation into Dr Clift's work. An internal inquiry had begun in 1977 when Dr Clift was moved to the HOFSS laboratory at Birmingham. During this inquiry shortcomings and errors in one of Dr Clift's cases were drawn to his attention and corrected. Other cases were then examined and 'found to be unsatisfactory'.[15] On the basis of these findings, the Home Office decided there was substantial prima-facie evidence that Dr Clift had selected results in writing his reports, had

[15] *Hansard*, 15 Nov. 1984, David Mellor.

reported results that were clearly wrong, and had been guilty of grave technical incompetence. On the basis of a report from Detective Chief Superintendent Provan Sharpe, charges were dropped by the Director of Public Prosecutions against three men accused of assault. Dr Clift's evidence concerning dust and bloodstains was considered crucial to the case. Given the private concern being expressed in forensic circles, the DPP decided not to proceed. The evidence itself was, therefore, never tested. A report prepared by Miss Pereira apparently criticized Dr Clift for failing to disclose the full facts, 'though she conceded that he could not have been responsible for a miscarriage of justice'.[16] No criminal intent was assigned to Dr Clift by the DPP.

In June 1978 three men (Morgan, Brown, and Cowley) were released from prison having served five- and six-year sentences for offences of burglary and robbery. Application for leave to appeal had, in the first instance, been refused. The case was referred by the Secretary of State to the Court of Appeal in 1978 'because fresh evidence cast doubt on scientific evidence which had resulted in the convictions'.[17] The Court of Appeal heard that there was virtually no identification of the robbers at the scene because they had worn masks. When the police had searched the suspects' house, they had found one pair of Adidas shoes and one pair of Europa shoes: 'As some foot impressions had been discovered near the scene of the crime Dr Alan Clift, a senior scientific officer at Harrogate, made a detailed consideration of the two pairs of shoes, with such assistance as the making of blocks and moulds. He concluded, and expressed his conclusion at the trial, that the condition of one of the shoes was that it was the shoe which made the mark in the soil.'[18]

The taking of plaster casts of footprint impressions is a standard forensic technique listed in numerous forensic science textbooks. At the original trial, the judge noted that Dr Clift had said that many minute irregularities were present on the heel, arch, and sole of the right shoe, corresponding to irregu-

[16] *Guardian*, 15 July 1981.
[17] *The Times*, 8 June 1978, and fieldwork notes of the case.
[18] Ibid.

larities in the plaster cast of the impression found at the scene of the crime. He concluded 'that the footwear impressions were made by the right Adidas shoe and not by any other'.[19] Dr Clift had also identified a pebble in one of the shoes which was said to correspond with a mark in the plaster cast. At the appeal evidence was led of a discussion which had taken place between Dr Clift and his colleagues in the laboratory at the time of the trial. One of these colleagues, Mr Kenneth Jones, recollected,

> I remember him stating that he thought there were sufficient characteristics of a unique nature for him to be able to say that one of the footwear marks left at the scene had been made by one of the training shoes and no other. We expostulated with him for some minutes with phrases such as, 'You can't say that, those marks are mould defects you would expect to find on all the shoes made by that machine.' When Dr Clift left, I felt that he had not been convinced by our arguments, but I did feel that he was going to look again. I heard later on that Dr Grant was coming to examine the exhibits on behalf of the defence, and that Dr Clift was re-examining the samples. Later on, probably three weeks or a month after this time, I heard that Dr Clift had found a pebble in one of the shoes which matched with a mark in one of the casts.[20]

During the course of the appeal, the court heard evidence from both Mr Jones and another expert, Mr Walker, that the mark was indeed consistent with having been made by the shoe. Mr Walker testified that, in his considered opinion, the right Adidas shoe could have made the mark shown on the cast but could not be uniquely connected with it. The point made by the Court of Appeal was that, from a legal point of view, there was a world of difference between saying that a mark was consistent with having been made by the shoe in question and saying that the mark was made by the shoe. The court decided that Dr Clift was going to the very limits by saying that the shoe was the very shoe worn by one of the robbers. The Court of Appeal concluded that, had the jury heard the arguments which had taken place within the four walls of

[19] Ibid. [20] Ibid.

the laboratory, it would have cast a doubt in their mind. Accordingly, the conviction became unsafe. It is clear, however, that the dispute would never have come to light had the convictions not been contested. There is no mechanism whereby such disputes can be brought out. The only mechanism available is the calling of witnesses and neither side has a duty to bring forth evidence which would undermine its case. It is doubtful, in any case, whether members of the scientific professions would wish to air professional disagreements before a non-scientific public. To do this would be to undercut their professional loyalties.

The Home Office asked the director of the Aldermaston laboratory, Miss Margaret Pereira, to produce an internal report on Dr Clift's work. Of some fifty cases examined, she concluded that in six cases Dr Clift had reached conclusions based on flimsy evidence, that he had failed to disclose the full facts, made risky decisions, reached unsound and invalid conclusions, written ambiguous reports, and shown general carelessness, inadequate scientific standards, and unbalanced reporting. One of these cases was *Preece*, in which, it was alleged, Dr Clift had given misleading evidence, first by claiming to be able to distinguish between mixed semen and vaginal stains, and secondly by failing to tell the court that Helen Wills had been of blood group A with a one in three chance of being a secretor. Several colleagues disagreed with the conclusions of the Pereira report. One senior forensic scientist, Dr Curry, concluded that although Dr Clift's evidence on semen grouping had been a gross error, his work on the *Preece* case did not show excessive zeal. Accordingly, Preece's solicitors should be informed that Miss Pereira's report showed no material scientific factual inaccuracies in Dr Clift's work. Dr Curry did not dismiss what Miss Pereira said in her report; his disagreement with her 'was merely a difference of emphasis . . . There are no ways in which you can quantify this.'[21] Other colleagues said that no case had been made out against him, since 'there is nothing in dispute which lies outside the bound-

[21] *Guardian*, 15 July 1981.

aries of professional opinion and judgment'.[22] There was, therefore, considerable peer support for Dr Clift.

However, a call was made by several Members of Parliament for a review of the *Preece* case, even though the then Home Secretary, William Whitelaw, had already issued several assurances that there was no cause for concern: Dr Clift had been suspended; his work did not affect anyone still in prison. The *Preece* appeal in 1981 undercut these assurances. In its wake, the Ombudsman, Sir Cecil Clothier, described Dr Clift's work as 'an unprecedented pollution of justice'.[23] The specific allegation was that Dr Clift had neglected to inform the court that Helen Wills had been of blood group A with a one in three chance of being a secretor. If this evidence had been heard, it was argued, the forensic link between the accused and the victim would have been fatally weakened. The jury might well have chosen not to convict. The High Court commented that, when he had given his evidence at the original trial, Dr Clift must have seemed 'a very impressive witness and there is no doubt that his evidence, supported as it was by Dr Gregory, was of crucial importance in providing for the jury sufficient evidence in law to support the verdict which they returned by a majority'.[24] The jury had to be satisfied that Dr Clift's evidence had been detached, and that it was 'wholly reliable and acceptable'.[25] 'At the time Dr Clift, highly qualified and experienced, appeared to be an expert witness as to whose quality, detachment and scientific reliability there was no doubt.'[26]

In fact, in his summing up to the jury the original trial judge, Lord Avonside, had been highly sceptical about the value of the expert evidence and did not encourage the jury to be persuaded by it. Only briefly reviewing the evidence about hairs, fibres, and paint, he made no mention at all of body fluids and stains on the victim's clothing: 'For [to] everyone

[22] Ibid.
[23] *Fourth Report of the Parliamentary Ombudsman*, Sir Cecil Clothier (HMSO, London, 1983–4).
[24] *Preece* v. *HM Advocate* (1981), Opinion of the High Court of Justiciary; also fieldwork notes of case.
[25] Ibid. [26] Ibid.

experienced in these Courts, it is known that if you get one expert to say yea you get another sure as fate to say nay: and that is why it is entirely, I am afraid, for you (ladies and gentlemen of the jury) to judge which forensic evidence you accept and which you do not.'[27] The defence had called its own forensic expert, who later said that he too had been aware that Helen Wills and John Preece were almost certainly both blood group A secretors. He had been surprised that defence counsel had not asked him to give evidence about this point.[28] However, in its review of the case, the High Court seemed to decide that Dr Clift alone had been aware that Helen Wills was also a blood group A secretor and that he alone was at fault for failing to disclose that knowledge to the court.[29] The court went on:

> We have no doubt that, consciously or unconsciously, Dr Clift expressed his confident opinion that the donor of the semen was an A secretor in a wholly misleading way. He did not disclose, in terms, that he was expressing this opinion as the result of tests made on mixed stains. He did not disclose Helen Wills' blood group or her probable secretor status when he knew (a) that it was impossible for the defence to discover this for themselves; and (b) that these were facts or matters which he now recognised were of 'enormous importance', which ought properly to have been disclosed to the jury, and which at once opened the door to what he called 'another interpretation' of the results of the tests . . . Every one of the distinguished scientists to whose evidence we shall come shortly was of opinion that the deceased's blood group and secretor status should have been disclosed to the defence and to the jury. In short, by giving his evidence as he did, he consciously or unconsciously reduced to a minimum the risk that his confident opinion that the donor of the semen was a blood group A secretor would be tested and, indeed, he deprived the defence and the Court of the means whereby it could have been called into question.[30]

Dr Clift's crime was that he failed to reveal gaps in the Crown's case. Nothing, however, was made of the fact that

[27] A. Brownlie, 'Expert Evidence in the Light of *Preece* v. *HM Advocate*', *Medicine Science Law*, 22/4 (1982), 237.

[28] *Hansard*, 15 Nov. 1984, Jack Ashley, 887.

[29] *Preece* v. *HM Advocate* (1981), Opinion of the High Court of Justiciary, 10.

[30] Ibid.

both the defence and the prosecution had failed to bring out the critical piece of information. No lawyer or expert is under any legal duty to reveal all the information in a case. Indeed, as Dr Clift himself is alleged to have pointed out, it is the law itself which shapes such practices because 'the Criminal Justice Act statement merely demands that what is said is true'.[31] There is no requirement that it be the whole truth.

Disclosure

Preece raises a particularly important issue for expert witnesses, that is, whether they are under any duty to reveal information to the court on their own initiative. Virtually everything we have looked at so far suggests quite the opposite. There is no legal duty upon the prosecution to disclose all its evidence. The revisions of the Police and Criminal Evidence Act 1984, which were supposed to improve the situation, say nothing about prosecution disclosure. The Police and Criminal Evidence Act 1984 confirmed the position adopted by the Royal Commission on Criminal Procedure. Even the strong view expressed in *R.* v. *Maguire* (1992) in fact places no duty on the expert to reveal evidence to the court on his own initiative.[32] In criminal cases the prosecution is under a duty to make a material witness available to the defence. This entails providing the names and addresses of any material witnesses, but there is no formal duty to disclose to the defence a copy of their statements. There may, therefore, be other evidence (including expert reports) about which the other side remains ignorant.

The Fisher Inquiry into the *Confait* case drew attention to the fact that the practice of the prosecution at that time was to leave the burden of requesting disclosure to the defence. Fisher criticized this practice because, in the absence of any knowledge about the kind of material available to the prosecution, the defence was in no position to make such requests. The net effect was that the defence never saw information which might

[31] M. Pereira, *New Scientist*, 1. Oct. 1981, 576.

[32] *R.* v. *Maguire* (1992).

be material to its case because it did not know it existed and could therefore not ask for it. Even if the defence had made such a request, it still had to rely upon prosecuting counsel to judge what would be material to a defence case with which he was not familiar.

As Fisher noted, there had in fact been a good deal of discussion on previous occasions about the practice of disclosure in criminal cases. It was an issue addressed by *Justice* in its 1966 report *Availability of Prosecution Evidence for the Defence.* It was also addressed by the James Committee, by the Devlin Committee,[33] and by section 48 of the Criminal Law Act 1977, which recognized the principle of entitlement to disclosure and contained a power to make rules providing for advanced disclosure. It was an issue for the appeal court in *R.* v. *Maguire* and has been much discussed in relation to procedure in civil cases. In 1973 the Solicitor-General promised to consider the matter,[34] and when Fisher was finishing his inquiry he noted that the Home Office was considering whether there were any defects in the system of disclosure. It was also a matter discussed by the Eleventh Report of the Criminal Law Revision Committee on Evidence.[35] In the Report of the Royal Commission on Criminal Procedure in 1981, which provided the initiative for an independent Crown Prosecution Service, the commissioners noted the disquiet aroused by *Confait* and the serious questions it raised about how the police handled the investigation. The case contributed directly to the setting up of the Commission in 1977.[36] The Commission discussed the issue of disclosure and concluded that openness was essential, but in summary cases the requirement for disclosure should only operate on request by the

[33] Report of the Interdepartmental Committee on *The Distribution of Criminal Business between the Crown Court and the Magistrates Court* (James Report), Cmnd. 6323 (HMSO, London, 1975); Report of the Departmental Committee on *Evidence of Identification in Criminal Cases* (Devlin Report), HC 338 (HMSO, London, 1976).

[34] Official Report HC 3 (Apr. 1973), vol. 874.

[35] Seventeenth Report of the Law Reform Committee on *Evidence of Opinion and Expert Evidence*, Cmnd. 4489 (HMSO, London, 1970); The Eleventh Report of the Criminal Law Revision Committee, Cmnd. 4991 (HMSO, London, 1972).

[36] Report of the Royal Commission on Criminal Procedure (Philips Report) (HMSO, London, 1981), para. 1.5.

defence. It further argued that in Crown Court proceedings the prosecution should but not must supply the defence with a copy of evidence which it proposes to call but which was not tendered at the committal proceedings. The defence should already have been supplied with copies of the evidence which the prosecution intended to call and which had already been tendered at committal proceedings.

This provision did not answer Fisher's criticism that the prosecution might well have available to it evidence which it did not intend to call, but which nevertheless might be material to the defence case. This was a problem in some of the IRA bombing trials in which the prosecution did not disclose material evidence to the defence. The Royal Commission Report was aware of this loophole in its proposals and the possibility that it might lead to miscarriages of justice. It referred to a working party set up by the Home Secretary in 1979 to look at this matter.[37] That working party had 'on the grounds of cost alone decided against any rule that the prosecution should as a matter of routine furnish copies of every single statement'.[38] The working party's preferred option was to leave the prosecutor with a discretion to disclose. Fisher's report had already noted the problem inherent in this option: it left the decision as to what was relevant or useful to the defence case with the prosecutor. The Royal Commission considered allowing the defence to ask the judge to order full disclosure but argued that for him to know what was relevant or useful to the defence case would first involve the defence disclosing its case. This, it argued, was inconsistent with the 'central feature of the accusatorial system that it is for the prosecution to prove guilt without assistance from the defence'.[39] Moreover, such a procedure would unduly tax scarce judicial manpower. The Commission concluded that the decision to disclose must remain part of the prosecutor's discretion. Thus the denial of disclosure to the defence was justified in paternalistic terms as protecting the defence from an

[37] Report of the Working Party on *Disclosure of Information in Trials on Indictment* (HMSO, London, 1979).

[38] Philips Report, 178. [39] Ibid.

unfair requirement to disclose its case. In fact, full disclosure (even accepting the Commission's reservations about sensitive material) would allow the defence itself to decide what is and what is not relevant without having to disclose its case. The Commission's rather weak solution was that the prosecution should disclose evidence having some bearing on the offences charged. There was to be no formal duty to disclose.

The Bar Code of Conduct advises prosecuting counsel to regard it as normal practice to disclose to the defence statements of persons he does not intend to call as witnesses, and to disclose statements by a prosecution witness which differ in a material respect from the evidence given by this witness in the committal papers. A decision in *R.* v. *Bryant and Dickson*[40] specified that the prosecution should supply to the defence particulars (names and addresses) of any witness who it knows can give material evidence but whom it does not propose to call itself. Another argument, raised by Lord Denning in *Dallison* v. *Caffrey* (1965)[41] was that the prosecution's duty went beyond the mere supplying of names and addresses to furnishing the defence with a copy of the statements of any such witnesses. This view was also taken by the court in *R.* v. *Lawson* (1989).[42] However, it is still up to defence lawyers to discover information which the prosecution has deliberately or inadvertently failed to disclose.

At first sight, a Practice Note issued by the Attorney-General in 1982[43] appears finally to settle this matter. It provided that all unused material should normally be made available to the defence. Unused material includes all witness statements not included in the committal bundle, statements of witnesses who are not to be called to give evidence at the committal; the unedited versions of any edited statements or composite statements included in the committal bundle. This would seem to be fairly exhaustive. However, there are provisos. First of all, this is only a guideline—what happens if the

40 *R.* v. *Bryant and Dickson* (1964) 110 JP 267 CCA.
41 *Dallison* v. *Caffrey* (1965) 1 QB 348.
42 *R.* v. *Lawson*, *The Times*, 21 June 1989.
43 A.-G.'s Practice Note (1982) 1 All ER 734.

prosecution fails to comply with it? Secondly, it only specifies that the material should *normally* be disclosed; there is no duty to disclose in all cases. *R.* v. *Maguire* makes it clear that failure to disclose would only constitute grounds for overturning a verdict where this amounted to a material irregularity. The Practice Note itself also specifies some circumstances when it would be better not to disclose. Moreover, prosecutorial discretion is even greater in triable either way and summary cases, where many of these provisions do not apply. In triable either way cases, the prosecution must disclose the statements of those witnesses upon whose evidence it will rely at trial, but it may choose instead to provide an outline summary of the facts which it proposes to adduce in the case. In summary cases this may be the best that one can obtain, and it is obtained only at the prosecutor's discretion. He may choose to disclose more than this, but he also may not.

In any event, it is unlikely that simply withholding information would in and of itself constitute grounds for overturning a verdict. To do this, one would need to show that the information withheld was material or relevant to the defence case, and to establish that its absence constituted a material irregularity in the conduct of the trial. It is difficult to establish this when one does not know exactly what material has been withheld; it is even more difficult to establish that the prosecution should have realized it was material to the defence case—after all, prosecutors cannot be expected to know what exactly the defence might construe as relevant and material. Furthermore, even though the law in the books may supply the safety net of an appeal procedure where these issues may be aired, it is by no means easy or usual for cases to gain access to the Appeal Courts. Yet, as becomes clear from the IRA bombing cases which I shall examine later in this chapter, the fact that information has been withheld by the prosecution frequently only comes to light during the course of an appeal on other grounds.

The law and procedure thus have a strong bearing on whether or not the accused will be able to discover what selections have been made by Crown experts. In *Preece* there was no legal duty on either expert witness to reveal the blood group and secretor status of Helen Wills. When the Crown failed to reveal this information, the defence lawyers were in a position to remedy the deficiency:

> Whatever was said in accusation of Dr Clift, the fact remains that neither the defence nor the prosecution, neither the trial judge nor the Appeal Court appear to have appreciated that the stains on the woman's knickers are likely to contain vaginal fluid and none asked about her secretor status—an error of omission which I would penalise heavily were it to be made by one of my LLB students in an oral examination. Like it or not, the function of a witness in our courts is to answer questions.[44]

The debate about this point prompted another writer, Sir David Napley, to comment that the true onus to reveal the omitted information lay squarely on the Crown.[45] In Pereira's view, however, 'It is incumbent on forensic scientists to be totally honest about their findings and not edit them in a way which might be prejudicial to either the prosecution or the defence.'[46] Yet pre-trial editing is a routine part of case construction. Indeed, Brownlie goes further and cites a legal text advising experts that in the United Kingdom they are under no obligation to reveal unfavourable information, or make out an opponent's case for him: 'It has always been thought sufficient if the witness gives his evidence fairly and moderately and leaves it to the opponent to draw attention to those aspects of the matter which are favourable to the opponent.'[47]

Even where there is a duty on the prosecution expert to disclose to the prosecuting authority (as in *R.* v. *Maguire*) there is still no duty on either the prosecution or the defence expert to disclose information to the court on their own initiative. Had the situation been different in *Preece*, the defence expert could have revealed to the court what he revealed some time later to the Press. The High Court itself acknowledged that in his

[44] Professor J. K. Mason, 43 *Bulletin of the Royal College of Pathologists*, 43 (1983), 3; J. K. Mason, 'Expert Evidence in the Adversarial System of Criminal Justice', *Medicine Science Law*, 26 (1986), 1; see also C. Goodwin Jones, 'Men of Science versus Men of Law: Some Notes on Recent Cases', in same volume.

[45] D. Napley, *Bulletin of the Royal College of Pathologists*, 45 (1984), 16.

[46] *New Scientist*, 1 Oct. 1981.

[47] Brownlie, 'Expert Evidence'.

initial report prepared for proceedings in England in 1973 Dr Clift had included the crucial fact that Helen Wills was of blood group A. In a later attempt to defend himself, Dr Clift also protested that he did disclose the victim's blood group. In 1984 he told a reporter: 'In fact I did disclose it. It is down in black and white beside me now, in the brief I prepared for counsel in the trial—"Blood from Helen Wills is of Group A".'[48] The original report had been made available to both sides at trial, but neither set of lawyers sought to lead this item of evidence. In 1984, a question in the House of Commons made the specific claim that Dr Clift omitted Helen Wills's blood group from the second report 'on the advice of prosecuting counsel'.[49] The same speaker also asked, 'If Helen Wills's blood group was so crucial that Dr Clift was discredited for not disclosing it, what blame can be attached to the defence for failing to probe this matter at the original trial?'[50] As with *Confait*, it was the expert not the lawyers who bore the full force of criticism.

Given all this, it is difficult to see, as Brownlie points out, why the High Court 'should have been so extremely critical of Dr Clift's evidence' and why they should have said that he deprived the defence of the means whereby his opinion could have been called into question.[51] Experts, as we have seen, are discouraged from volunteering information, as are all witnesses. It is thus difficult to see how the High Court could have concluded that Dr Clift fell far below the standards of accuracy, fairness, and objectivity which are *normally* to be expected and normally displayed by such a witness.[52] Every expert witness is told to confine himself to the questions he is asked, and is very often sharply reminded that it is not for him to volunteer evidence but simply to answer the questions. Mason has pointed out that the evidence of experts, like that

[48] *Mail on Sunday*, 11 Nov. 1984.

[49] R. S. Ormrod, Transcript of the 33rd Kettle Memorial Lecture, 19 Apr. 1982; see also R. S. Ormrod, 'Evidence and Proof: Legal and Scientific', *Medicine Science Law*, 12 (1972), 19.

[50] Ormrod, 33rd Kettle Memorial Lecture.

[51] Brownlie, 'Expert Evidence'.

[52] See *Preece* v. *HM Advocate* (1981), Opinion of the High Court of Justiciary.

of other witnesses, is taken not given.[53] There is no rule of evidence which says that an expert has a duty to speak up when he knows he is being inadequately questioned.

The Expert's Duty?

Even if we allow that experts should speak up when they realize they are not being adequately examined, we are still left with the problem of what happens if the expert does not realize that he is being inadequately examined. On what basis can the man of science judge the forensic skills of a man of law? As a witness in a court of law, he is supposed to defer to the law's procedures for truth finding. Further, what if the expert has been previously advised that certain parts of his evidence are irrelevant, inadmissible, or immaterial? How can the man of science overrule the advice of his lawyer? How can he know which evidence is material, admissible, and ought to be disclosed unless he is also steeped in the law? The experiences of Dr Clift and Dr Cameron suggest that he is damned if he does and damned if he doesn't.

A further issue which arises here is the question of what happens when the forensic scientist discovers something in the course of his investigations which he was not requested to explore. Lawyers have a general permission to edit out of experts' reports any irrelevancies; comments about phenomena which fall outside those which the expert was originally asked to investigate could be said to fall into this category. Does the expert ignore such findings simply because the lawyers advise that they lie outside his remit? He may be told that it is irrelevant to the Crown case but it may be highly relevant to the defence case. If it is not in his report, it will never come to light.

Sir Roger Ormrod has insisted that experts bring attention to the margin of error in their scientific conclusions and the limits on the inferences which can properly be drawn from the scientific observations. He has argued that in a scientific setting

[53] Mason, 'Expert Evidence in the Adversarial System'.

(e.g. reading or publishing a paper) it is standard practice to define the limits of accuracy of the experimental results; in presenting results to a lay audience it is necessary to go further in many instances and to point out what the results do *not* prove.[54] Yet HOFSS practice may also have deterred Dr Clift from including in his report anything of which he was not entirely certain. There was some suggestion, for example, that the oral swab administered to the victim provided no definite indication of her blood group. Some laboratories have a policy of leaving out of reports any inconclusive test results. In other words, they fail to tell us about the selections they have made in the course of their investigations. The practice was brought to the notice of the Home Affairs Committee by Henry Bland, who described some techniques of report writing which down-graded evidence contrary to the Crown case:

> there is no doubt a strong tendency which appears to have grown up over the years for some laboratories and their scientists to denigrate any negative evidence which they obtain; it would appear this is done so as not to destroy the police case. To give some idea of the criticism I make. Often a blood stain cannot be grouped, there are many reasons why a blood stain may not be grouped, one of them being it is very old and in the context of the particular matter this might be a significant point. The scientist's report will either not make any reference to this or may go so far as to say 'no significance should be attached to the failure to obtain a grouping in this case', leaving the defence to scrabble around to look for a possible explanation which might be of vital input to their client's case. Similar comments limiting the damage of a negative report are more frequently found in connection with document cases where handwriting identification is weak.[55]

Bland added, 'There is a tendency to denigrate what I term negative evidence, that is evidence where it does not help to convict a person but it could help the defence to say it was not this man in the box. There is a tendency to say that because it has not worked or does not show a link it may not matter in

[54] Ormrod, 'Evidence and Proof'.

[55] HAC Report on *The Forensic Science Service* (HMSO, London, 1989), vol. ii, Minutes of Evidence, Henry Bland, 169.

this case for this reason or that reason.'[56] Moreover, a prosecution expert may not realize the relevance or significance of a piece of information. When his terms of reference have been supplied by the prosecution, he may fail to appreciate the implications for the defence side. Quite often, even where experts include such information in their reports, the lawyer in charge of the case may decide not to produce that particular report in evidence. The information need not, then, necessarily be disclosed. The selections made in the writing of reports are only part of the total number of selections.

In her work in the sociology of science, Karen Knorr-Cetina has noted the following: 'The constructive operations with which we have associated scientific work can be defined as the sum total of selections designed to transform the subjective into the objective, the unbelievable into the believed, the fabricated into the finding, and the painstakingly constructed into the objective scientific fact.'[57] What is actually required of the forensic scientist is exactly this making of selections, the transformation of the subjective into the objective. The problem arises if this is publicly explicated, since it opens the expert up to comparison with the empiricist image of scientific endeavour, an image which he himself has helped to perpetuate.

Post-Preece

In the aftermath of *Preece*, Dr Clift was retired on the grounds of 'limited efficiency' and the Home Office undertook an investigation of at least 1,500 other cases in which he had carried out the forensic investigation. The investigation was felt to be necessary because otherwise confidence in expert witnesses would be badly shaken, the task of the courts would become harder, and a disservice would be done to the forensic science service.[58] Sir Cecil Clothier commented that it was evidence of the excellence of the HOFSS that trust in forensic science had

[56] HAC Report on *The Forensic Science Service* (HMSO, London, 1989), vol. ii, Minutes of Evidence, Henry Bland, 169.

[57] K. Knorr-Cetina, 'The Ethnographic Study of Scientific Work', in K. Knorr-Cetina and M. Mulkay, *Science Observed* (Sage, London, 1983), 121.

[58] *New Law Journal*, 131 (Jan.–Dec. 1981), 666.

become so profound: 'But the depth of that trust is a measure also of the extent of disaster when it is found to have been betrayed.'[59] The issue had thus become one of public confidence in the forensic science service generally. It was seen as imperative that this be restored. Although the Home Office claimed that there was insufficient evidence at the time to suggest that Dr Clift's findings were tainted across the board, it eventually reviewed 129 cases in detail, sending sixteen cases to the Court of Appeal for review. These were cases in which the accused had pleaded not guilty but had been convicted. (It is unclear what happened to those cases in which the accused had pleaded guilty.)

In total, the cases selected for review spanned fourteen years and included six convictions for murder, one for rape, one for indecent assault, two for robbery, one for attempted buggery, one for incest and indecent assault, one for aggravated burglary, one for affray, one for actual bodily harm, and one for manslaughter. Contrary to previous Home Office assurances, in some of these cases the accused were still in prison. The investigations included an examination of Dr Clift's laboratory notebooks for each case, though some material for 1970 and 1971 had not been available because police records had been routinely destroyed. The Court of Appeal reviewed the cases in 1984. For a variety of reasons not of all them were in fact heard. Four of those convicted did not wish to have the matter reopened and one could not be traced. Eleven cases were eventually sent for review. In three the appellant subsequently abandoned the appeal, and in another four convictions were overturned. The other appeals were dismissed.[60] In those appeals which were dismissed, Dr Clift's evidence at the original trial was upheld by other forensic experts. The Court of Appeal regarded it as being accurate and in no way wrong.[61]

In a written answer to Jack Ashley, MP, in November 1984, the Home Secretary pointed out that, in those cases where the

[59] Fourth Annual Report of the Parliamentary Ombudsman, para. 54.

[60] The convictions which were overturned were those against Szptyma, Sample, Gilfellan, and Mycock (fieldwork notes).

[61] Fieldwork notes of observed Appeal Court cases, 1984.

conviction had been quashed, 'I understand that in no case did the court state that it had quashed the conviction solely on the ground that Dr Clift had handled the material.'[62] He also stated,

> It has come as no surprise to me that in a majority of them the Court of Appeal has felt that the convictions were perfectly safe, or the appellants have abandoned their appeals . . . But we should have been much criticised had we failed to expose to open view those cases where a contrary view was genuinely possible even if not probable. This is what we have done. It has been a painful exercise for all of us at the Home Office and particularly for the Forensic Science service. But I am convinced that what we have done in the wake of the Preece decision was both wholly necessary and quite right. These matters have been fully ventilated and appropriate procedures have been followed.'[63]

The Home Office was criticized by Lord Lane, the then Lord Chief Justice, who commented in two cases that 'it would not have made a h'porth of difference whether Dr Clift . . . had given evidence or not'.[64] He went on to say, 'We beg leave to doubt whether sufficient consideration was given as to which cases should have been referred to us and those which should not have been.'[65]

Other Discredited Experts

The Clift cases demonstrate (1) the impact which advocacy may have upon the expert's input into legal proceedings; and (2) the means by which an expert's work may be deconstructed long after it was first completed. The process of investigation into the Clift cases began with the assumption that any miscarriage of justice must be due to the shortcomings of Dr Clift as an individual. It proceeded to look for examples of his prosecution-mindedness and, where selections were found, these were taken as proof positive of his bias. There is a long list of very eminent expert witnesses who have suffered a similar fate. Dr Alfred Swaine Taylor, one of the founding fathers of foren-

[62] *Hansard*, 15 Nov. 1984, Jack Ashley, 303.
[63] Ibid. 893.
[64] *Scotsman*, 19 July 1984.
[65] Fieldwork notes.

sic science, suffered stinging criticism when, in the trial of Dr Smethurst for the poisoning of Isabella Bankes in 1859, he was forced to concede that the traces of arsenic he had detected could equally have come from contaminated laboratory equipment. Mr Letheby's evidence at the trial of William Palmer was undermined by his admission that, in an earlier trial, he had made a mistake. Dr Bronte and Dr Camps also suffered from the same accusations of bias. Even Spilsbury suffered a similar allegation in the *Thorne* case, but since his credibility was absolutely essential to the Crown, it was imperative that he be rehabilitated. He could 'persuade a jury to accept a bad scientific proposition more readily than others could achieve acceptance of a right view'.[66]

The list of discredited experts is extensive. After *Confait*, Professor Cameron was profoundly criticized in the *Chamberlain* case. Dr Donald Nelson was criticized by a Royal Commission in New Zealand for allowing 'personal vanity and a stubborn determination to be right at all costs . . . colour his evidence'.[67] Dr Keith Gugan was discredited in an English trial for lacking scientific objectivity and for clouded judgement.[68] In 1981, the dismissal of another HOFSS scientist, Colin Horncastle, gave rise to more adverse publicity about the service. Horncastle was said to be a deranged dissident who wrote what other forensic scientists regarded as a bizarre paper on toxicology.[69] Other forensic scientists disagreed with this judgement, however. Nevertheless, Horncastle was removed from casework and his competence to submit evidence was called into doubt. The Home Office later sought to retire him on the grounds of mental illness. In 1984 the Home Office was also forced to admit that a prosecution expert had given scientific evidence in a murder trial when he was not qualified to do so. The evidence, given by John McCafferty, played a crucial part in securing the conviction of Paul Cleeland in 1973. It concerned a chemical test for traces of lead on Cleeland's

[66] J. Shaw, 'The Law and the Expert Witness', *Proceedings of the Royal Medical Society*, 69 (1976), 83–9.

[67] A. Brownlie, 'How often are Experts Right?', in A. Brownlie (ed.), *Crime Investigation* (Scottish Academic Press, Edinburgh, 1984).

[68] Ibid.

[69] 17 *Medicine Science Law*, 17 (1977), 37.

clothes, possibly from the discharge of a shotgun. However, it transpired that McCafferty was not qualified to give this evidence, his expertise lying in the field of firearms, not chemistry. *The Times* called for a retrial on the basis that McCafferty was not competent to give scientific evidence and that tests had been available at the time of trial which could have distinguished between lead from the environment and lead from a shotgun. Since Cleeland was a painter and decorator, this test could have been of some significance. In 1990 and 1991 more HOFSS scientists (Dr Higgs, Dr Elliott, and Dr Skuse) were severely criticized. It is to their cases that we now turn.

The IRA Bombing Cases

The three cases with which I am concerned here are colloquially known as The Guildford Four, the Maguire Seven, and the Birmingham Six.[70] All three cases stemmed from IRA bombing incidents on mainland Britain during the 1970s. They demonstrate empirically the ways in which guilt is constructed out of the public view in the pre-trial stages of prosecution. The Court of Appeal's review of the case against the Guildford Four and the Birmingham Six, together with Sir John May's inquiry into the convictions of the Maguire Seven, provide insights into the way in which forensic findings are selectively employed to produce strong Crown cases. Such scrutiny provides us with a rare glimpse of the process of case construction behind the scenes. It is rare precisely because, contrary to the rhetoric of the adversary system, Crown cases are seldom challenged in open court, seldom thoroughly tested

[70] The Birmingham Six originally comprised nine defendants: *R.* v. *Wiliam Power, Hugh Callaghan, Patrick Joseph Hill, Robert Gerald Hunter, Noel Richard McIllkenny, John Walker, James Kelly, Michael Bernard Sheehan, and Michael Joseph Murray*, 16 June 1975; six appealed: *R.* v. *McIllkenny, Hill, Power, Walker, Hunter, and Callaghan* (1988) 86 Cr. App. R. 181–6; (1989) 88 Cr. App. R. 40–8; *R.* v. *McIllkenny and Others* (1991) 93 Cr. App. R. 287; (1992) (CA) 2 All ER 417. The Maguire Seven: *R.* v. *Anne Rita Maguire, Patrick Joseph Maguire, William John Smyth, Vincent John Patrick Maguire, Patrick Joseph Paul Maguire, Patrick Joseph O'Neill, Patrick Joseph Conlon* (1976); (1992) 94 Cr. App. R. 133; *R.* v. Maguire (1992) (CA) 2 WLR 767; (1992) (CA) 2 All ER 433; see also a Report into *The Convictions Arising out of the Bomb Attacks at Guildford and Woolwich* (May Inquiry) (HMSO, London, 1990). The Guildford Four: *R.* v. *Hill*, *R.* v. *Conlon*, *R.* v. *Armstrong*, *R.* v. *Richardson* (1974, 1975) The Times, 20 Oct. 1989.

at appeal. If they were, we would discover that selectivity is the norm rather than the exception.

The Guildford Four

The selections made in the case of the Guildford Four first became apparent at the trial of the group known as the Balcombe Street Four.[71] Four people had already been convicted of the bombings at Guildford and Woolwich but the Balcombe Street group claimed that they had been responsible for these bombings. Work carried out by a principal scientific officer at the Royal Arsenal Research and Development Establishment (RARDE) suggested that the same group had indeed been responsible for all the bombings in and around London, including the Guildford and Woolwich bombs. When cross-examined, he stated that all the bombings clearly conformed to the same pattern. Although he had originally produced a report which showed these links, his later report did not mention the Guildford and Woolwich bombs. He stated in evidence that the reason for this omission was that an officer from the Bomb Squad had told him not to include the Guildford and Woolwich bombs. Another forensic expert consulted by the police had also examined the series of bombings and under cross-examination stated that the Guildford and Woolwich bombs could well have been connected with the others carried out by the Balcombe Street group. The claims of the Balcombe Street group threw some doubt on the convictions of the Guildford Four. The belief that the right people had already been convicted led the police and the prosecution to dismiss this piece of disconfirming evidence. As with *Confait* and *Preece*, a strong case had been constructed; any evidence which threw it into doubt must be wrong. The forensic scientists in this instance were guided by police instructions. Their

[71] This group comprised IRA members held at a siege at Balcombe Street in London. It was claimed that they were the main IRA active service unit operating in mainland Britain at the time of the Guildford and Woolwich explosions, and that they were in fact also responsible for these attacks.

disconfirming evidence was edited out and never disclosed to the defence.

The Maguire Seven

Following a confession by one of the Guildford Four, Gerard Conlon, which implicated his 'Aunt Annie' as a bomb-maker, the Maguire family was arrested in December 1974. After forensic tests were carried out, all defendants were accused of having kneaded and manipulated explosives. No bulk of explosives nor traces of explosives were found in their homes. The only evidence against the accused came from tests designed to detect whether they had been handling the explosive nitroglycerine. The tests were of swabs taken from the palms of their hands and scrapings taken from underneath their fingernails. These were analysed at RARDE, using Thin Layer Chromatography (TLC). The tests proved positive for the presence of nitro-glycerine on the hands of six of the seven accused. At the trial, the Crown asked the jury to infer from the traces detected by the tests that the defendants had indeed handled explosives in bulk. It also insisted that the scientific tests were 'like fingerprints'; they conclusively proved that the defendants had been handling explosives. In particular, a finding of traces of nitro-glycerine under the fingernails ruled out the possibility of 'innocent contamination'. Doubts about the reliability of the tests were in fact raised at trial by the former director of RARDE, Dr John Yallop, the man who devised the TLC test. He argued that 'no competent scientist could do other than conclude that the hypothesis is incorrect; namely that the pink spot is not due to nitro-glycerine. To do otherwise would be unscientific, illogical and pig-headed.'[72] Dr Yallop agreed that he had himself testified on previous occasions that the TLC test was specific for nitro-glycerine, but after leaving RARDE he had changed his mind. In his view, the results could have been caused by another substance, referred to at trial as 'Substance X'. The defence also con-

[72] B. Woffinden, *Miscarriages of Justice* (Hodder & Stoughton, London, 1987), 263.

tended that the nitrates detected in these tests could as easily have come from household cleaning agents or tobacco smoke.

Dr Yallop appeared as an expert witness for the defence. Under cross-examination he was accused by Sir Michael Havers, QC, of not having been honest and frank with the court, of having selected results which best supported his evidence and having discarded the rest: 'An expert witness such as yourself has the obligation to be frank with the court . . . not to be selective about his experiments . . . not to pick the best and discard the worst . . . not to select the ones that suit the case you were supporting and discard the one that casts doubt upon it . . . You have not followed good scientific practice by disclosing all the tests; you have just been selective and picked out the one you wanted.'[73] As the May Inquiry later revealed, if anyone had been selective about their scientific evidence, it was the prosecution rather than the defence experts. Sir John May's Report in 1990 questioned the reliability, fairness, and credibility of the prosecution's expert evidence which, he said, had been fundamental to the convictions of the Maguire Seven.[74] Their evidence was challenged on precisely the same grounds that Sir Michael Havers had challenged Dr Yallop. At the original trial it was Dr Yallop who had raised the issue of the specificity of the tests used to detect nitroglycerine. He discovered a document which had been sent to him by one of the prosecution experts, Mr Elliott. In this document, dated June 1974, Mr Elliott specifically stated that the test used did not discriminate between nitro-glycerine and another type of substance, pentaerythritol tetranitrate (PETN). This document became known as Exhibit 60. It posed difficulties not only because of its contents but because Dr Yallop had discovered it after all the evidence had been heard and the judge was about to start his summing up. In order to produce the document, the defence would have to recall Dr Yallop, and the prosecution would probably want to recall Mr Elliott. This was problematic because the jury had already heard the evidence in the case as well as counsels' closing speeches.

[73] Cited by the May Inquiry, 51. [74] Ibid. 50.

The compromise reached by the lawyers was that both sides would meet out of court to discuss the implications of the document. Defence lawyers raised the question of why the Crown experts had not mentioned the potential confusion between nitro-glycerine and PETN in their evidence at trial. Their suggestion was that, in failing to bring out this point, the Crown experts had misled the court. The judge implied that Dr Yallop must also have done so.[75] During these discussions between counsel and the judge, the second Crown expert, Dr Higgs, apparently spoke to Mr Elliott by telephone, and then informed prosecuting counsel that for PETN to have produced the same results, 'there would have had to have been a wholly unrealistic concentration of PETN' on the test.[76] As Sir John May was later to point out, this information reassured prosecuting counsel that they need not worry about PETN 'and could regard it as a red herring.'[77] Mr Higgs noted that Sir Michael Havers told the court that 'his evidence would show that the experts could show that the substance on the swabs was not PETN'.[78]

The defence considered recalling both the Crown experts and Dr Yallop in order to challenge the reliability of the scientific evidence but decided against it. Dr Yallop's cross-examination by Sir Michael Havers had been 'an effective challenge . . . not only against his scientific expertise but against his credibility generally'.[79] The defence was therefore reluctant to expose Dr Yallop to a second ordeal. They eventually agreed to a compromise solution, which was to put Exhibit 60 before the court without any further oral evidence, but simply with 'a short agreed explanatory statement'.[80] Sir John May commented that 'defence counsel clearly expected the judge to make the point to the jury that the defence relied on Exhibit 60 as showing that the test was not unique . . . [the judge] took the point . . . In the event, however, the judge did not put this point to the jury.'[81] He told the jury that it could ignore Exhibit 60 and the idea that the substance shown on the swabs was PETN. In fact, as Sir John May later pointed out, Exhibit 60 removed the main plank in the Crown's case.

[75] Cited by the May Inquiry, 62. [76] Ibid. 32. [77] Ibid. 33.
[78] Ibid. 34. [79] Ibid. 26. [80] Ibid. 34. [81] Ibid.

In the event, all accused were found guilty of possessing nitroglycerine for an unlawful object.

The application for leave to appeal in 1977 focused on the accuracy and reliability of the TLC tests. The Court of Appeal decided that the scientific evidence was clear and of considerable weight. It refused leave to appeal. Three years later the Home Secretary again refused to review it on the grounds that no new evidence had been brought forward. Technically, this was correct, for the doubts about the accuracy of the tests had already been raised at trial. Scientists in the HOFSS had been aware of but had suppressed these doubts. In a 1984 television programme an independent forensic scientist, Dr Brian Caddy, assessed the scientific evidence. He concluded that the TLC test was unreliable and that it failed to provide sufficient evidence that the compound detected had been nitro-glycerine.[82] In 1985 the case was again raised in the House of Lords but again the Home Secretary found there were no new grounds for reviewing the case. In the late 1980s, however, doubts concerning the confession evidence of the Guildford Four resulted in a review, and in the setting up of the May Inquiry to examine the circumstances surrounding the convictions. The fragile and interpretative nature of the scientific findings began to be revealed for the first time. Sir John May concluded that he could not rely upon the RARDE tests, and that any attempt to do so would be impracticable and wholly unreliable. The results of the hand swab tests were also brought into question. The forensic scientist, Mr Elliott, had concluded in 1974 that these tests indicated that 'an explosive substance has been handled recently'.[83] However, in 1990 Sir John May noted first that no description of his testing procedures 'let alone detailed results' were included in Mr Elliott's statement. His second observation was that the positive results obtained by Mr Elliott's tests were of a scale which was 'something of a rarity in the laboratory'. Mr Elliott went on: 'There was a great deal of excitement. Never before had we seen so many positives on a plate at a reasonably high level of intensity. We just did not

[82] Ibid. 34. [83] May Inquiry, 22.

believe it, quite honestly.'[84] Clearly the scientists were greatly surprised that the facts had spoken for themselves in quite such an unequivocal fashion.

Investigations were carried out for the May Inquiry by Professor Thorburn Burns. He concluded that the TLC test could not possibly distinguish between PETN and nitroglycerine. The conclusion drawn by May was that there was far greater scope for innocent contamination than the evidence of the Crown witnesses at trial had suggested: 'In particular, the assertion by Mr Elliott that each male defendant must have manipulated or kneaded a primary source of contamination, that is a quantity or bulk of explosive, is not borne out by subsequent investigations.'[85] Mr Elliott's assertions had been made in the context of the form sent to the laboratory, which recorded that the Maguires were suspected of bombing offences. Sir John May commented that this led the scientists to set a wider parameter during their tests than they would otherwise have done. The parameters regarded as safe by the scientists during their tests were thus consequential. The context in which materials came into the laboratory affected the type of tests they carried out, how they were carried out, and what interpretation was put upon them. The claims made by Mr Elliott were never fully tested at trial, partly because the defence lacked access to adequate forensic facilities. Furthermore, at the trial only extracts from notebooks of the scientists who gave evidence were introduced as exhibits. By contrast, the May Inquiry managed to secure all the notebooks and copies of documents from RARDE. At the trial, requests by the defence for the notes were refused on the grounds of irrelevance. A note by Mr Purnell, QC, part of the prosecuting team at the trial, included the phrase, 'Defence seek in advance of trial bench notes of Elliott. Prosecution hedge.'

It appears that the prosecution experts never revealed the results of disconfirming tests to the prosecuting lawyers. It also appears that the prosecuting lawyers never asked them about such tests, nor did they discuss the scientists' notes with them at

[84] May Inquiry, 23. [85] Ibid. 29–30.

the time. When they were later criticized for this, they argued that 'Counsel are entitled to rely on the expert's accuracy and the completeness of his account . . . There was nothing which I saw or heard which could have put me on notice that we were being told less than the whole picture.'[86] Far from being irrelevant to the case, when the notebooks were eventually disclosed to Sir John May he found that they revealed 'a number of disturbing features'. For example: 'It is amply demonstrated by the RARDE notebooks that not only did the RARDE scientists know throughout the trial that PETN was potentially confusable with nitro-glycerine on the TLC test in toluene, but also that their work at RARDE had taken this fully into account.'[87] This had not been mentioned by any of the RARDE scientists at the trial, despite the fact that RARDE had carried out specific tests to try to distinguish between nitro-glycerine and PETN. According to Sir John May, 'it is hard to believe that PETN was far from their minds'.[88] He also stated that he was satisfied that Mr Higgs did have PETN in mind but did not mention it, and described Mr Higgs's testimony at the trial as 'less than frank'.[89] Mr Higgs failed to mention that PETN was not distinguishable from nitro-glycerine during either of his two pre-trial conferences with counsel. His explanation was that 'It was never raised in any of the previous trials as being in conflict with nitro-glycerine and, of course, we had decided that it just did not feature in the present IRA campaign as a possible explosive.'[90]

This brings home an earlier point. The remit of the forensic scientists was set by the prosecution. Since the cases were considered in the context of IRA bombings, PETN was thought to be irrelevant; it fell outside the scientists' remit. Sir John May later commented that he felt it was 'improper for the scientists to presume in that way to exclude it'. He also argued that after a while Mr Higgs deliberately left PETN out of the list of substances which could mimic nitro-glycerine:

It had become clear to Mr Elliott and Mr Higgs following the consultation in July 1975, if not before, that the Crown's case rested on

[86] *Independent*, 19 June 1990. [87] May Inquiry, 37. [88] Ibid. 38.
[89] Ibid. 39. [90] Ibid. 38.

the specificity of the TLC test using toluene to identify nitro-glycerine. They knew that the test was not specific for nitro-glycerine when advising the prosecution team and when giving evidence, but they failed to say so. They knew that a second system was available to resolve the two substances but they did not mention it. Whilst these failures were in my view deliberate, I do not believe they were borne of a conspiracy to deny justice to the defendants. The scientists had honestly reported positive results for nitro-glycerine on 4th December 1974. Once charges were laid and the defendants were committed for trial, there was no going back on what had been said. The scientists wrongly believed that they could rationalise their exclusion of PETN. They imperfectly understood their duties as forensic scientists and as witnesses.[91]

This begs the question, what exactly are the duties of a forensic scientist? How do they relate to his duties as a witness? Is the expert under any duty to disclose unsolicited information which may damage his side's case? Where, in law, are the duties of a forensic scientist clearly spelt out so that they may not be so frequently imperfectly understood? When the Court of Appeal looked at *R.* v. *Maguire* in 1992 it only specified that prosecution scientists should disclose evidence from all their tests to the prosecuting authority. The appellants argued that the scientists had failed to reveal all the facts which were relevant, and that they were selective about the experimental data which they produced, 'choosing material which supported their case while discarding that which did not'.[92] This is exactly the practice identified by Henry Bland in his evidence to the Home Affairs Committee; it is exactly what might be expected of HOFSS laboratories with a policy not to disclose the results of tests which seem uncertain; it is exactly what lawyers try to ensure experts will do. As with *Confait* and *Preece*, however, it was the experts and not the lawyers who were criticized.

Moreover, without the disclosure of the laboratory notebooks these criticisms could not have been made and it is far from routine for the prosecution to disclose the laboratory notebooks of its experts. How, then, could one normally know

[91] May Inquiry, 47. [92] *R.* v. *Maguire* (1992).

whether an expert had misunderstood his role as a scientist and as a witness? It further transpired from the RARDE notebooks that a second set of tests had been carried out by Mr Elliott, on the hand swabs and gloves, using a different TLC system. These second tests were negative. They were never disclosed to the trial court. Sir John May noted: 'Both Mr Higgs and Dr Hayes denied knowledge of these second tests until they examined the papers at RARDE shortly before my public hearing began. Even then, when they first gave evidence to me, they did not tell me of the second tests.'[93] The laboratory notebooks revealed that, despite the fact that a specific test to discriminate between PETN and nitro-glycerine was in regular use at the time of the original trial, the court was led to believe that no further testing was possible. Mr Higgs maintained this position in a note for a meeting at the Home Office in 1983, and at his first appearance before the May Inquiry. The notebooks also showed that other tests had in fact been carried out during the Maguire trial, but that their results were only partially disclosed. One of these tests established that crushed heart tablets could also produce a positive result for nitro-glycerine.

The failure to communicate the results of these tests was laid by Sir John May firmly at the door of RARDE scientists. This was despite the fact that the scientist carrying out the tests 'could only have been asked to do them by a senior scientist at RARDE, and one who knew what questions were being raised at trial'.[94] This again raises the issue of whether there is in fact any duty on the expert to reveal to the court the limits of his findings. Even when the defence had asked for the notebooks to be disclosed, the prosecution simply used a perfectly legitimate legal procedure to deny access. Later, Sir John May was to state that this was regrettable and significant. Had the prosecution 'been more open minded to disclosure of the scientists' notebooks it is very possible that events would have taken a different turn. The respective responsibilities of counsel and of those who instruct them are general matters of

[93] May Inquiry, 40. [94] Ibid. 43.

importance with which I propose to deal at a later stage of this Inquiry.'[95] As regrettable and significant as this approach may have been, it was clearly permitted by legal procedure. This was not, therefore, a case of the law in practice breaking the law in the books; it was simply a case of prosecutors using the law in the books to help secure a conviction.

Sir John May argued at the end of his interim report that the position on disclosure in 1990 differed substantially from the position in 1975. It had, he argued, been altered by the Attorney-General's Guidelines in 1982 and the introduction of the Crown Court (Advance Notice of Expert Evidence) Rules 1987. The introduction of the Police and Criminal Evidence Act (PACE) in 1984 is also frequently cited as a guarantee that miscarriages of justice such as the Maguire case could never happen again. This view is over-optimistic.[96] PACE is silent on prosecution disclosure. *R.* v. *Maguire*, far from improving the situation, emphasizes the prosecuting expert's role as part of the prosecuting team, with a duty to disclose to the prosecuting authority but not directly to the court. It is still up to the prosecution to decide what might be material to the defence. None of this guarantees full and frank disclosure. In any event, even in an all-cards-on-the-table approach to disclosure the prosecution is under a duty to reveal only that case which it intends to rely upon at trial. The remaining nine-tenths of the iceberg may contain material relevant to the defence but how is the defence to know? What sanctions operate if the prosecution fails to abide by ethical and judicial guidelines about disclosure? As in *Preece*, even where crucial evidence has been disclosed to the prosecution, there is no guarantee that the prosecuting authority will bring it out at trial. Moreover, in those cases where the accused pleads guilty, the prosecution case is never tested at trial. The fragility of prosecution forensic evidence is therefore never exposed.

[95] May Inquiry, 48.

[96] See M. McConville, A. Sanders and R. Leng, *The Case for the Prosecution* (Routledge, London, 1991).

The Birmingham Six

Following the Guildford Four and Maguire Seven cases a further appeal against conviction arose in the case of the Birmingham Six in 1991. The case was sent back to the Court of Appeal by the Home Secretary's Reference following a protracted public campaign waged on the issues of (1) the accuracy of the scientific evidence; and (2) the accuracy of the confession evidence. It was alleged that the scientific evidence was unreliable and that the confession evidence had been fabricated and/or extracted through police brutality. The six accused were tried in June 1975 for bomb attacks in the city of Birmingham. Five of them were arrested *en route* from Birmingham by train to the ferry to Ireland from Heysham. They had been playing cards on the train, a factor which was later adduced to question the accuracy of the scientific tests. Following their arrest at Heysham, the suspects were taken to a police station where they were subjected to forensic tests carried out by Dr Frank Skuse, a forensic scientist from the HOFSS laboratory at Chorley. He tested the hands of the five suspects for traces of nitro-glycerine using three different tests.

On one of these tests, the Griess test, he obtained positive results for Hill's right hand and Power's right hand. All the other swabs tested were negative. Using the more sensitive TLC test and Gas Chromatography, Dr Skuse obtained a negative reaction for Power, a negative result for Hill's right hand, but a positive result for Hill's left hand. Testing for another component of explosives, ammonium nitrate, he obtained positive results for Hill, Power, Walker, and himself.[97] The tests from Hunter and McIllkenny proved negative throughout, as did those on Callaghan, who had been arrested in the West Midlands and was tested at Sutton Coldfield. On the results of the Griess test Dr Skuse declared that he was 99 per cent certain that Hill and Power had been handling explosives. On the results of the TLC and Gas Chromatography tests he pronounced himself 100 per cent certain.[98] The Griess test has

[97] Woffinden, *Miscarriages of Justice*, 285.
[98] *R.* v. *McIllkenny* (CA) (1991) 93 Cr. App. R. 296.

since fallen into disuse, though its drawbacks were already known to the forensic scientists and to the Crown at the time of the original trial. They were not made known either to the defence or, at a later date, to the Court of Appeal. As in the Maguire case, the defence at trial did attempt to introduce doubts about the reliability of the tests used. Dr Black, an independent consultant, formerly HM Chief Inspector of Explosives for the Home Office, appeared as the defence expert witness. His main criticism was that the tests used could not discriminate between nitro-glycerine, ammonium nitrate, and nitro-cellulose. Nitro-cellulose may be used in the production of paints and varnishes; one may pick it up from shiny surfaces such as public house furniture, soap, patent shoes, and playing cards. Dr Black's contention was that the tests used could equally well show positive for varnishes, fungicides, insecticides, and petrol additives found in the soil as well as in the atmosphere.[99] In other words, a positive test result could equally imply contamination by nitro-cellulose commonly used in everyday substances by the majority of the population. The tests could not, therefore, be relied upon as having isolated and identified nitro-glycerine.

Dr Black also testified that the Griess test should only be regarded as a preliminary test since he did not regard it as specific for nitro-glycerine. His views on the Gas Chromatography test were not heard since Dr Skuse was not himself questioned on this issue: 'In particular, he was about to give evidence, which might have been important, about certain graphs or traces produced by the GCMS equipment. But Dr Skuse had not been asked about the traces. So they were never produced, and the point was allowed to drop. In Dr Black's view the GCMS result . . . did not prove conclusively the presence of nitroglycerine on Hill's left hand.'[100] The Court of Appeal eventually acknowledged in 1991 that it was the trial judge who had failed to explain to the jury the difference in view between Dr Skuse and Dr Black:

[99] Woffinden, *Miscarriages of Justice*, 285.

[100] *R.* v. *McIlkenny* (1991) 297.

As I have already said, this is such a complicated subject that I am not going to try and reproduce the technical difference on this issue. If I start trying to explain to you the mysteries of gas chromatography and mass spectrometry, what you see and what you do not see with an oscilloscope, and what is meant by atomic units and blips, and so on, I shall be getting hopelessly out of my depth and, I suspect, out of yours too, members of the jury. There it is. Was the result described by Dr Skuse about this scientific question as obtained by the GCMS test used by him using the Hill swab from the left hand, or was it not? Alternatively, viewed from another angle, is it as complex as that?'[101]

He went on to advise the jury that if it decided that it preferred Dr Black's views to those of Dr Skuse,

then you will obviously conclude that the forensic evidence of Dr Skuse is of no value. Indeed, Dr Black's theory logically seems to imply not only that Dr Skuse's theories were of no value, but that Dr Skuse has been spending and must have spent much of his professional life wasting his time because, if Dr Black is right, the Griess test was not worth carrying out . . . Do you think Dr Skuse has been wasting most of his professional time? It is a matter entirely for you.[102]

Dr Black was upset 'because I thought the evidence I gave was not being given the weight that it merited . . . Too often at these trials the word of the prosecution scientist is taken as gospel.'[103] At the conclusion of the trial, the accused were found guilty on 126 murder charges and sentenced to life imprisonment. The trial judge commented that they had been convicted 'on the clearest and most overwhelming evidence I have ever heard in a case of murder'.[104]

The seams in the Crown case began to unravel many years later, after a prolonged effort by the accused's lawyers and campaigning groups. This required the defence to secure a rare resource, independent forensic experts. The Court of Appeal in 1991 commented upon the disadvantage faced by the accused in criminal trials in the following terms: 'Since the jury cannot embark on a judicial investigation the material

[101] Ibid. [102] Ibid. [103] *Observer*, 27 Oct. 1985.
[104] Woffinden, *Miscarriages of Justice*, 294.

must be placed before them. The great advantage of the adversarial system is that it enables the defendant to test the prosecution case in open court, but its disadvantage may be that the parties are not evenly matched in resources, particularly when it comes to the carrying out of expensive experiments to test the scientific evidence.'[105] As we have already seen, the advantage of the Crown in respect of scientific expertise is guaranteed by the prevailing structure of forensic services in England and Wales. As far as the Court of Appeal was concerned this criticism was met by the obligation on the part of the Crown to make available to the defence material which may prove helpful to the defence case. This obligation often amounts to no more than lip-service. The point is brought home by the fact that, during the course of the final Birmingham Six appeal, it was alleged that the DPP's office had known at the previous appeal in 1987 that scientific evidence existed which threw doubt upon the convictions. This had been withheld from the defence and from the Appeal Court.

The initial application for leave to appeal against the convictions had been dismissed in 1976. An action against the police for assault eventually came before the Court of Appeal in 1980. The allegation was that the confession evidence had been obtained by the use of force and violence. In his judgment on this matter, Lord Denning commented:

> If they won, it would mean that the police were guilty of perjury; that they were guilty of violence and threats; and that the confessions were involuntary and improperly admitted in evidence; and that the convictions were erroneous. That would mean that the Home Secretary would either have to recommend that they be pardoned or to remit the case to the Court of Appeal . . . That was such an appalling vista that every sensible person would say, 'It cannot be right that these actions should go any further'. They should be struck out either on the ground that the men are estopped from challenging the decision of Mr Justice Bridge, or alternatively that it is an abuse of the process of the court. Whichever it is, the actions should be stopped.[106]

[105] *The Times*, 28 Mar. 1991.

[106] Lord Denning is reported to have made these comments in 1980, during an

A similar argument was raised in Australia against further reviews of the so-called 'Dingo Baby' case in which Lindy Chamberlain had been convicted of the murder of her daughter Azaria:

> Many greeted the decision to hold an inquiry into the Chamberlains' convictions with comments such as 'It's too expensive,' or 'She's had a trial and two appeals—you've got to stop somewhere'. Even now when it is plain that they should not have been convicted, some suggest that we may have paid too high a price for justice. It is not merely a matter of money but of loss of confidence in the system. There seems to be a feeling that the myth of infallibility must be maintained at all costs. If that means occasionally convicting the innocent then they should presumably be content with the knowledge that they have been martyrs in a good cause.[107]

At the end of the Birmingham Six action against the police, Lord Denning concluded that 'the cases showed what a civilised country we are: through the facility of legal aid, we had allowed the six Irishmen a generous crack of the judicial whip.'[108]

The appellants eventually secured a referral back to the Court of Appeal in 1987. The appeal was the result of new evidence introduced by the MP Chris Mullin, which focused on alleged misconduct on the part of the investigating officers. In addition, apart from a particular problem with the specificity of the test used to detect nitro-glycerine, the trace used in the Gas Chromatography Mass Spectrometry tests was later said to have been too small to be identified definitively as nitro-glycerine. The results obtained by these tests could also have been obtained from other substances, including chemicals found on the hands of smokers. An HOFSS scientist responsible for quality assurance commented that the results of these tests in Hill's case were not acceptable as a test for nitro-glycerine.[109] Moreover, tests which had been carried out on

appeal by the West Midlands Police against a joint civil action by the accused for injuries allegedly sustained whilst in police custody. Lord Denning's decision not to allow the case to proceed was later upheld by the Law Lords in 1981.

[107] K. Crispin, *The Dingo Baby Case* (Lion Publishing, Tring, 1987), 354.

[108] Woffinden, *Miscarriages of Justice*, 35.

[109] Dr. Alan Scaplehorn, *Daily Telegraph*, 15 Mar. 1991.

two other men travelling to Ireland by train and boat on the same evening as the accused had also proved positive. These suspects had been handling shiny adhesive tape. Dr Bamford's evidence concerning the results of these tests was not made available to the defence. Indeed, the existence of these tests was not disclosed by the prosecution until 19 October 1987, immediately prior to the appeal. Two independent experts, Dr Brian Caddy, of the Forensic Science Unit, Strathclyde University, and David Baldock, former senior scientific officer with the Home Office, both confirmed the lack of specificity for nitro-glycerine of the Griess test. Other substances would give the same results.

After a lengthy hearing the appeal was dismissed in 1988. The Lord Chief Justice, Lord Lane commented, 'The longer this case has gone on, the more the court has been convinced that the jury was correct.'[110] Public disquiet coupled with fresh scientific evidence secured a further appeal in 1991. Evidence showed that since a positive result in the scientific tests could be obtained from soap, there was a possibility that traces of detergent on the porcelain bowls used by Dr Skuse to conduct the tests could have been a source of contamination. Tests carried out by a team headed by Dr Scaplehorn confirmed that detergent used by Dr Skuse to wash out porcelain bowls could have caused the tests to give a positive result.

At the time doubts about the forensic tests were first raised, the Home Secretary regarded the defence scientists' view (that the test used was not specific for nitro-glycerine) as wrong because it was based upon an incorrect formula supplied to the defence experts by the director of the Chorley laboratory. Details of the precise formula used by Dr Skuse had been requested by David Baldock in 1985. On one version of the test given by Dr Skuse, the method he used would have failed to distinguish between nitro-cellulose and nitro-glycerine; on the second version of the test, there would have had to have been so much nitro-glycerine on the hands of the accused that the police would have been able to see it. In fact no nitro-

[110] *R.* v. *Callaghan and Others* (1988) 86 Cr. App. R. 181–6.

glycerine had been seen on the hands of the two men. The Home Office later said that it had failed to find any records of the exact procedures used by Dr Skuse. At the appeal, Dr Skuse is reported to have claimed that he did not keep any records of his experiments: 'I didn't deem it necessary because I was happy with my materials and my methods.'[111] However, records were located amongst the papers kept by Dr Black, who had consulted with Dr Skuse at the time of the original trial. These records clearly indicated the use of a 1 per cent solution. Dr Caddy contended that this concentration would fail to discriminate between nitro-glycerine and nitro-cellulose. Like Dr Clift, Dr Skuse was dismissed from the HOFSS on the grounds of 'limited efficiency'.

Failure properly to contest the expert evidence is a fundamental flaw of a system whose strength is supposed to lie in contest. Not only may the unreliability of a particular piece of evidence fail to come to light, but the interpretative nature of scientific work is itself glossed over. Just as there is nothing in scientific practice which encourages the public to see how science works, so there is nothing in law which compels the Crown to make an appeal hearing a full rehearing of all the evidence. The Birmingham Six appeal is a case in point. In making significant decisions about how to process a case, practitioners are guided by the structure of law, by its rules and regulations. Their actions are thus frequently not an abuse of due process but a use of due process. In this instance, the Director of Public Prosecutions and prosecuting counsel acted, as the court noted, 'with perfect propriety'. The Director of Public Prosecution's decision not to oppose the appeals limited any damage which might have been caused by full revelation of the details of the case. (A similar decision in one of the Clift cases meant that there was no opportunity for the appellant or the experts to air their views.) When the Home Secretary sent the Birmingham Six case back to the Court of Appeal, the Director of Public Prosecutions decided not to contest the scientific evidence raised by the appellants. Indeed, the DPP's

[111] *New Scientist*, 26 Nov. 1987.

office did not wish to contest the appeal at all. The Court of Appeal itself, however, felt it should not give its secret blessing to an agreement arrived at between the defence and the prosecution. The appeal was therefore heard and fresh scientific evidence was submitted with regard both to the reliability of original tests and to the accuracy of the confessions. At the conclusion of the 1991 appeal, the convictions were finally found to be unsafe and unsatisfactory.

As the final appeal of the Birmingham Six ended, in March 1991, plans were already well under way for the HOFSS to become an executive agency from April 1991. This new status was heralded as a significant step away from close ties with the prosecution and police investigations. In fact, the revised system greatly increases HOFSS dependency on the police and threatens to effect a State closure on all forensic science services. It also significantly reduces the likelihood of detecting the kinds of selections made in the cases discussed here.

11 Conclusion

In 1839, Lord John Russell remarked that 'the people of the country in general . . . feel that there is a power and supremacy in the law, to which they are ready to yield obedience'.[1] In fact the nineteenth century saw a number of challenges to the rule of law, not least from the new scientific experts, who offered an alternative foundation for social life based upon the natural order of things. The co-optation of science by law gave judicial decisions, essentially matters of evaluation and choice, an air of finality. Instead of being cultural artefacts, as durable and as contestable as any other, legal decisions became the inexorable and immutable working out of a natural or given order.

I have tried to show how expert witnesses have contributed to the production of this facticity. The legal process encourages selectivity amongst experts in their man of law role, whilst at the same time punishing them for such selections when they fail to live up to an idealized version of the man of science. There is no rule of evidence which says experts have an overriding duty to the court; all we have are one or two notable comments from the judiciary. On the other hand, there is a plethora of rules and procedures which specifically say that experts are called by the parties as witnesses on their behalf. The law of evidence makes it abundantly clear that experts have no independent *locus*. Getting rid of this was, after all, one of the benefits of moving experts into the witness-box in the first place.

Experts have also become victims of their own past performances. By continually recreating the image of an impartial man of science every time they enter the witness-box they have constructed the idealized standard by which future experts will be judged. Caught between the unrealistic

[1] Russell, cited in D. Philips, *Crime and Authority in Victorian England* (Croom Helm, London, 1977), 283.

expectations of a man of science and the *realpolitik* of law, the expert witness occupies a particularly fragile position. This is made even weaker by the fact that the expert's status is linked to that most artificial of legal distinctions, the separation between fact and opinion. Since experts are only useful if they help lawyers win cases, lawyers must also ensure that they retain a certain credit with the courts. There is, then, an in-built tension between the lawyer's wish to discredit an expert in a particular case and his wish to preserve the overall currency of expert evidence. The imperatives of advocacy coalesce with the professional interests of scientists to act as a check on courtroom excesses which would undo their overall credibility. However, the inherent weakness of their individual position as witnesses means that the collective credit of experts has always been tenuous. The cases I have discussed here expose this fragility. Scientists are constantly at risk of being hoist by their own positivist petard.

The expert witness is one location where there has been an interpenetration of law and science. This may also be seen in the growing scientization of law and judicialization of science. In the nineteenth and twentieth centuries, lawyers have looked to science for examples of rational truth-finding procedures. Increasingly, scientists also look to legal modes of resolution as a model for settling their own debates. What may once have seemed to be hard-and-fast distinctions between science as a body of knowledge and law as a body of knowledge have become blurred as the practices of science and law each intrude one upon the other. Law is often the arbiter of scientific practice. It decides whether some kinds of practice are or are not ethical, who is and who is not a real as opposed to a bogus expert, what is and is not real science. Policy-relevant scientific debates are also often forced into the adversarial model of public inquiries as in the law courts. In both settings, science is generally a secondary player in the sense that it is in the witness-box rather than on the Bench.

One of the common features of the miscarriage of justice cases in the 1980s and 1990s has been the pillorying of experts whose scientific work has been blamed for wrongful conviction.

There has followed a temporary unease about the criminal justice system, but since it is the scientists who are to blame, the overall currency of law is retrieved. The distinctions between law and science which structure the debate thus serve to maintain the overall dominance of law. Cumulatively, these so-called miscarriage of justice cases have led us to rethink the place of expert evidence in the legal system. Yet the implications are also probably more profound than is generally realized. The search for a remedy to our problems with expert evidence proceeds along disappointingly predictable lines. Thus, for example, it has been suggested the problems would disappear if a panel of experts replaced the jury in complex cases, if judges became more specialized, if they sat with scientific assessors, if they were replaced altogether with science courts. I have shown that the advantages and drawbacks of each system have already been debated at length. Independent court experts have been tried, tested, and rejected by a judiciary anxious to protect its own power; science courts offer more rather than less opportunity for dispute. Nevertheless, as long as society believes that science delivers truth it will confine itself to the goal of finding the perfect system of divination.

The debate about expert witnesses has become atrophied by this discourse. Lawyers in the English legal system look to Continental court expert systems for a solution, whilst lawyers on the Continent look to England and America for remedies to their problems. Neither side has found what it is looking for. This is because the certainty they seek is a human construct of relatively recent vintage, a beguiling shibboleth. The belief in the power of science to convict the guilty has become so entrenched in society that it is difficult to shift, but just at the point where we are beginning to acknowledge the contestable nature of legal verdicts, science has come up with a miracle cure—we are confidently assured that DNA fingerprinting offers incontrovertible proof of guilt. The philosopher's stone could have no greater transformative power.

The role of the expert represents a particular moment of articulation between law and science in the modern world.

Together they powerfully affect social life, combining to produce seemingly unassailable judicial verdicts and policy outcomes. They promise to transform the world by delivering the benevolent progress of modernity. Where there is inequality and subjectivity they promise equality and objectivity; where there is doubt they promise certainty. Looking behind the scenes at the construction of cases allows us to see what kind of certainty and what version of the truth is constructed by the prevailing practices of law and science. We can see how the black and white, cut-and-dried, self-evident reality of the judicial verdict is accomplished and how alternative versions are silenced.[2] Whether science is actually capable of delivering truth and certainty is less important than the fact that it is generally believed to be able to do so: it is this belief which is the critical resource of advocacy. Scientists have encouraged the concomitant belief that there is an unambiguous truth which men of science alone can perceive. Since there can only be one truth, men of science must agree on what it is—consensus not conflict will therefore be the norm. This belief in the consensual nature of science has given lawyers a stick with which to beat the experts, but it has also provided the experts with a convenient means by which they may be spared the spectre of a plural science. In the past, both science and law have mobilized a rhetoric of positivism to structure and maintain their interests, organization, power, and activity.[3] They have done so with far-ranging social consequences,[4] one of which has been a tendency to organize our 'kaleidoscopic flux of impressions'[5] into what Gellner calls a conceptual unification of the world.[6] For Gellner, science needs one world and its specific historical achievement has been its creation of such a world. But this view no longer convinces; increasingly, it fails to command popular as well as academic support. Moreover, as

[2] D. McBarnet, *Conviction* (Macmillan, Oxford, 1981).

[3] C. Bazerman, introd. to C. Bazerman and J. Paradis (eds.), *Historical and Contemporary Studies of Writing in Professional Communities* (University of Wisconsin Press, Madison, Wis. 1991), 3–10.

[4] Ibid.

[5] B. L. Whorf, *Language, Thought and Reality* (MIT Press, Boston, 1954), 213.

[6] E. Gellner, 'Relativism and Universals', in M. Hollis and S. Lukes (eds.), *Rationality and Relativism* (Blackwell, Oxford, 1982), 200.

Bazerman points out, the world 'cannot be reduced to the rhetorical domination of a powerful monolithic discourse of science and technology . . . professional discourses may hold much influence over many aspects of our lives, [but] they provide varied enough voices to maintain a robust rhetorical environment and keep the forces of reductionism at bay'.[7]

Once scientific and legal facts can be shown to be negotiated constructs, what Wynne has termed the fragile achievement of collective interaction, negotiation and social consensus is revealed.[8] Science itself is reconstituted as a social practice, 'a set of customs, social institutions and personnel, sets of books, papers, documents and apparatus, and a great many errors, side-tracks, complete nonsense. This is real science and this is also rational science.[9] None of this need be fatal to the social utility of science and law, though it is fatal to what Wynne calls the *ex cathedra* character of their pronouncements. It is also fatal to treasured legal concepts such as causation and the fact/opinion distinction.

Understanding science and law as social processes does not mean that we reject their possibilities and achievements;[10] it does mean we see more clearly what interests they serve, what kind of order they reproduce, and regard both as essentially contestable. Scientists may still provide us with versions of reality which work for some (but not all) practical purposes. These purposes are, however, shaped by our socio-economic and ideological interests. Instead of being reifications of a particular social order, science and law may become transformative social practices. Recognizing the social and cultural nature of science and law entails recognizing that they have long upheld specific sorts of social orders: 'whatever the moral and political values and interests responsible for selecting problems, theories, methods, and interpretations of research, they reappear at the other

[7] Bazerman, introd.

[8] B. Wynne, 'Science and Law as Conflict-Resolving Institutions: The Implications of Recent Sociology of Science', paper for the annual meeting of the British Society for the Philosophy of Science, University of Durham, 1982.

[9] I.C. Jarvie, *Rationality and Relativism* (Routledge, London, 1984).

[10] P. Bell, 'Science and Tort: Towards a Sickening Imbalance', paper for the Annual Meeting of Law and Society, Amsterdam, June 1991.

end of the enquiry as the moral and political universe that science projects as natural and thereby helps to legitimate. In this respect, science is no different from the proverbial description of computers: "junk in, junk out".'[11] Once this is acknowledged, we may still allow the lawyers and scientists to fashion with us (not for us) some raft of stability upon which social actors can depend, though it may be of a very different complexion from the one to which we presently cling. In Harding's view, it is within the moral and political discourses that we should expect to find paradigms of rational discourse, not in scientific discourses claiming to have disavowed morals and politics.[12]

Judicial verdicts and their scientific underpinnings are essentially contestable. In the means they use to construct their cases lawyers and scientists routinely conceal this. Trial records do not allow us to see how cases have been constructed. In order to discover this, we have to go backwards through the process unpicking how cases have been pieced together pre-trial. This has been the method adopted by inquiries investigating possible miscarriages of justice. My own observations, based on watching lawyers work upon cases *in situ*, have given us an added insight into the ways in which cases are constructed. Revealing the constructed nature of judicial verdicts entails acknowledging their essential contestability. Where we place the limits on this contestability is not a foregone conclusion fashioned for us by the nature of the facts themselves. It is a matter of 'local consensus',[13] of choice about where we set the parameters of intelligibility, certainty, and justice. To say this is also to suggest a social programme for sociology of law and sociology of science. Without their input, discussions about the nature of the legal system and the social uses of scientific images will remain the privilege of a political and legal élite. This is where the debate about the criminal justice system is currently located. It reminds us that some persons have greater

[11] S. Harding, *The Science Question in Feminism* (Open University Press, Milton Keynes, 1986), 251.

[12] Ibid.

[13] M. Hesse, *Revolutions and Reconstructions in the Philosophy of Science* (Harvester, Hassocks, 1980), 4350.

access to debates than others, a greater say in where the parameters are set, greater control over what choices are made and what kind of order prevails. For local consensus, therefore, we must at present read the consensus of certain powerful groups. The social order which judicial verdicts legitimize is an order which enables some people (and not others) to explain, control, predict, and manipulate their environment.[14] One issue, therefore, is who controls and who is controlled, who manipulates and who is manipulated?

The rational organization of everyday life by science and law is part of the grand narrative of modernity. In the practical as well as the ideological transformation of the modern world, experts have become especially powerful agents, revealing and maintaining a particular social order. Science promised to change the world and emancipate humankind from ignorance, poverty, and oppression. It promised it in this world rather than the next. In courts of law, however, science proved a useful agent of legitimization for the oppressions of the prevailing social, economic, and political order. Despite the grand egalitarian claims of law and science, the existing order in court generally supports and reflects a system of settled inequalities associated with conservatism and oppression. This order is constructed both in detail and in grand design, its givenness is routinely achieved by harnessing science to law. It is not a natural order divined because the facts speak for themselves. It is the particular expression of prevailing social interests, devoid of any higher claims to reason, certainty, or truth.

The Rationality versus Relativism Debate

Debunking the positivist view of law and science, showing that they are social activities, is not to deny that science and law are sometimes useful ways of doing things. It is to argue that this is all they are. Our definitions of usefulness are culturally and temporally defined. Problematically, the constructivist (or

[14] S. Lukes, 'Relativism in its Place', in Hollis and Lukes (eds.), *Rationality and Relativism*, 297.

deconstructivist) approach to law and science has led some to contend that those engaged in looking at the social construction of reality are actually engaged in its destruction.[15] The sociologist's deconstruction of what Hollis has termed the gold standard of truth will have confirmed, for some at least, their belief that sociologists are a pretty shifty lot. What began as an attempt to understand the socially constructed nature of truth, certainty, and rationality has resulted in a critique which has cut the ground from under our feet. Facts are no longer independent, the old dichotomies of nature/culture, fact/opinion, objective/subjective, truth/falsity are all undone. It seems that all that remains is rampant subjectivism and relativism. Since for the constructionists there never was any independent objective gold standard of truth this is less worrying to the non-believers than to the believers. The old certainties were never certain; they too were shaped by particular ideological and material formations. To say this is to say nothing very new. Sociologists of science have attacked the notion that scientific knowledge ought to be uncritically granted a privileged status, arguing that it is 'those who oppose relativism . . . who pose the real threat to a scientific understanding of knowledge and cognition'.[16] To their critique post-modernism has added fundamental questions not only about the possibility of progress, certainty, and truth but about their desirability. Elsewhere in the arts and humanities, this has contributed to the death of the idea of universal truth and the embracing, instead, of the possibility of diverse realities. Truth and reason have become relativized: 'For the relativist there is no sense attached to the idea that some standards or beliefs are really rational as distinct from merely locally accepted as such.'[17] On this view, truth and rationality are cultural products with a particular socio-historiography. Their ordering of the cosmos must be understood in terms of the social and material conditions in which they developed and the culture in which they make sense. Our understanding of the cultural authority of science

[15] M. Hollis, 'The Social Destruction of Reality', in Hollis and Lukes (eds.), *Rationality and Relativism*, 67–8.

[16] Hollis and Lukes (eds.), *Rationality and Relativism*, Introd.

[17] Ibid.

and law must therefore proceed by placing them within the history of ideas and institutions:

> the incidence of all beliefs without exception calls for empirical investigation and must be accounted for by finding the specific, local causes of this credibility. This means that the sociologist . . . must search for the causes of [a belief's] credibility. In all cases he will ask, for instance, if a belief is part of the routine cognitive and technical competences handed down from generation to generation. Is it enjoined by the authorities of the society? Is it transmitted by established institutions of socialisation or supported by accepted agencies of social control? Is it bound up with patterns of vested interest? Does it have a role in furthering shared goals, whether political or technical, or both? What are the practical and immediate consequences of particular judgements that are made with respect to the belief?[18]

Sociology of science and sociology of law have thus helped deconstruct their subjects; this book has its place in this effort. However, this is unlikely to comfort those who seek a programme of reconstruction. Some hard questions remain about how decision-making may be carried on in our society. Because the legitimacy of judicial and scientific findings resides so much in their claim to universality, truth, and certainty, once the role of judicial and scientific creativity is revealed their social authority seems irrevocably to slip away. To argue this is to argue that if objectivity and neutrality fail, science and law become systems of subjectivity, partiality, and politics, discourses which render up a particular (rather than a universal) world-view. This is to imagine, of course, that they rendered up universalism in the first place. The universality of law and science was always a raft of our own (or someone else's) making. As Popper argues, the empirical basis of objective science

> has nothing 'absolute' about it. Science does not rest upon a solid bedrock. The bold structure of its theories rises, as it were, above a swamp. It is like a building erected on piles. The piles are driven down from above, into the swamp, but not down to any 'natural' or

[18] B. Barnes and D. Bloor, 'Relativism, Rationalism and the Sociology of Knowledge', in Hollis and Lukes (eds.), *Rationality and Relativism*, 22.

'given' base; if we stop driving the piles deeper, it is not because we have reached firm ground. We simply stop when we are satisfied that the piles are firm enough to carry the structure at least for the time being.[19]

This is what science and law do. The fear of a descent into the doldrums of anarchic relativism is part of the ideological buttressing of the prevailing social order which warns that to abandon our present certainties is to let loose the Owl of Minerva.

If the reign of positivist law and empiricist science is irrevocably over, if all the assumptions, hopes, and pursuits of modernity are misguided, what then remains? The answer would seem to be some form of cognitive and cultural relativism. However, recognizing our gold standard of truth as a cultural product does not of necessity entail some slippery slope into moral or normative relativism.[20] If the old science and the old law were, as I have suggested, also systems of choice, then the new science and the new law will likewise be systems of choice—but different choices. Where we set our parameters, what we regard as certain for all practical or transformative purposes, is just such a matter of choice. Such choices require the unity of knowledge combining, as Harding puts it, moral and political with empirical understanding: 'It sees inquiry as comprising not just the mechanical observation of nature and others but the intervention of political and moral illumination.'[21] In place of the universal and objective, sociologists, historians, and others now positively seek to recover the subjective. In place of the universal they seek to understand the particular; in place of one world-view they prefer a plethora of world-views. The interpretative constitution of fact has been restored, blurring again the line between fact and opinion with which the story of expert witnesses began. The debate has been joined. Whether we like it or not, it will not go away.

[19] K. Popper, *The Logic of Scientific Discovery* (3rd edn., Hutchinson, London, 1962).

[20] G. Evans (ed.), *Asia's Cultural Mosaic: An Anthropological Introduction* (Prentice Hall, Singapore, 1992).

[21] Harding, *The Science Question*, 241.

Index